DISCOVERING COMMUNICATIONS

DISCOVERING
COMMUNICATIONS
by John Stansell

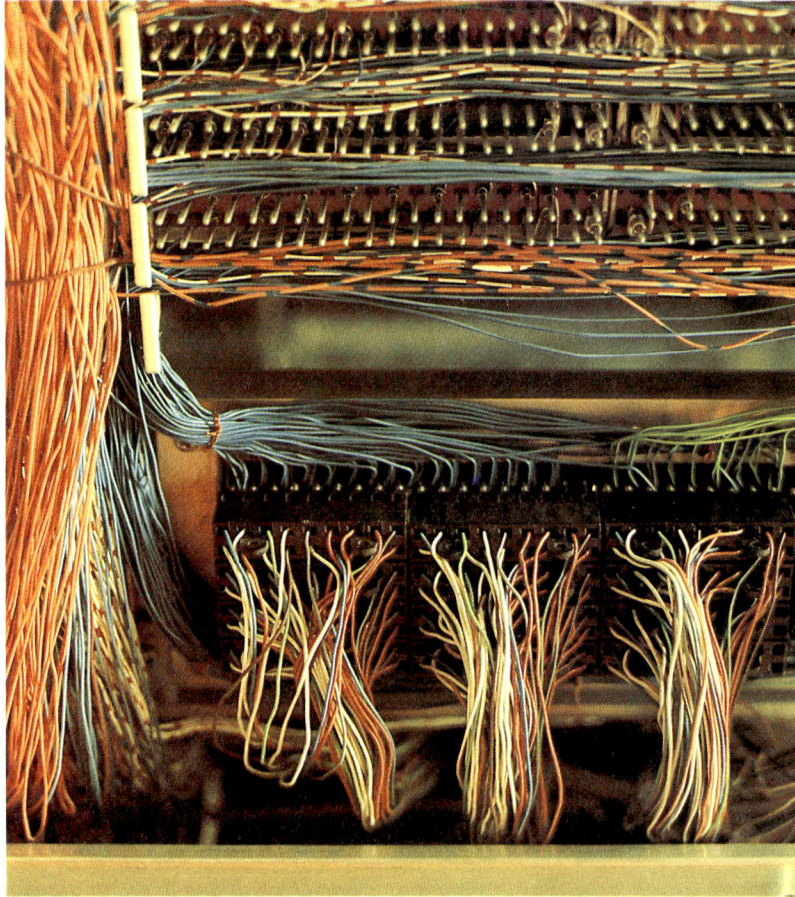

Published by

T

STONEHENGE

in association with

The American Museum of Natural History

The author
John Stansell is Technology Editor for the popular magazine *New Scientist*, an international weekly science and technology journal. After being trained as an electrical engineering officer in the British Royal Navy, he began his career in journalism in 1970 with a weekly magazine serving the electrical engineering industry, specializing in research and development. He lives in England with his wife and daughter.

The consultant
Martin W. Nabut, a member of the administrative group at Bell Laboratories, has had more than 20 years of experience as a writer and editor of science and engineering publications. He has been managing editor of the Bell Laboratories *Record* and has written for trade and technical publications on topics ranging from electronic switching systems to laser fusion, the latter during his five years as director of communications at the University of Rochester's Laboratory for Laser Energetics.

The American Museum of Natural History
Stonehenge Press wishes to extend particular thanks to Dr. Thomas D. Nicholson, Director of the Museum, and Mr. David D. Ryus, Vice President, for their counsel and assistance in creating this volume.

Stonehenge Press Inc.:
Publisher: John Canova
Editor: Ezra Bowen
Deputy Editor: Carolyn Tasker

Sceptre Books Ltd.
Editorial Consultant: James Clark
Managing Editor: Barbara Horn

Created, designed and produced by
Sceptre Books Ltd., London

© Sceptre Books Ltd., 1982

Library of Congress Card Number: 82-50628
Printed in U.S.A. by Rand McNally & Co.
First printing

ISBN 0 86706 013-1
ISBN 0 86706 064-6 (lib. bdg.)
ISBN 0 86706 033-6 (retail ed.)

Set in Rockwell Light by
Facet Filmsetting Limited, Southend-on-Sea, England.
Separation by Adroit, Birmingham, England

Contents

The History of Communications	6
Basics of Communications	8
The Telephone	10
Telephone Switching I	12
Telephone Switching II	14
The Telephone System	16
Cables	18
Facsimile	20
Telex and Supertelex	22
Radio I	24
Radio II	26
Antennas	28
Amateur and Mobile Radio	30
Microwave	32
Secure Communications	34
Radar	36
Sonar	38
Civil Satellites	40
Military Satellites	42
Satellites as Tools	44
Communicating with Light	46
Fiber Optics	48
Integrated Optics	50
Traditional Electronics	52
Electronics Today	54
Black and White Television	56
Color Television	58
Flat Screen Television	60
The Role of the Computer	62
Viewdata and Teletext	64
Digital Communications: Voice	66
Digital Communications: Data	68
Recording Sound: Records	70
Recording Sound: Tape	72
Traditional Hi-Fi	74
Stereo and Surround Sound	76
Digital Hi-Fi	78
Video: Tape	80
Video: Disk	82
Network Systems	84
Electronic Office	86
Voice Recognition	88
Telemetry	90
The Incredible Future	92
Glossary	94
Index, Credits and Bibliography	96

The World of Communications

Just imagine the world of today without telephones, radios, or televisions. We use these and many other means of communicating constantly and depend on them to an extraordinary degree, but how many people understand, for example, how sound waves are able to travel through wires to someone thousands of miles away; how a picture magically appears on a television screen; or how an airplane descending from 30,000 feet can land on a strip a few miles long?

Communications is an ever-growing field of knowledge that embraces not only man's natural means of exchanging information, such as language, but the technological means as well — telephony, computer technology, and satellite communications.

From the simple to the complex, the machines we use today function because of natural and physical laws. Without a knowledge or understanding of these laws and the technologies built on them, it is impossible to appreciate the miraculous progress made in communications over the past century.

This book helps the reader to understand the essentials behind the devices currently used in communications. Starting with basic concepts — such as what electricity is and how it works, and how radio waves travel — it moves through the major areas of communications that are so important today and will be in the future — the electronic office, voice recognition, and digital communications.

The historical aspect of communications — the astonishing rate at which it has progressed since the middle of the last century and the people who have made it all happen — is another fascinating area. How, for example, the vacuum tube technology in use since the turn of the century all but disappeared in less than two decades, to be replaced by the solid-state revolution, profoundly illustrates the speed at which communications has and will continue to grow.

Everyone will be affected by the discoveries and decisions being made at this very moment in laboratories around the world. We will not have to wait long before the ideas and dreams of today become practical realities. Currently, the entire field of telephone technology is being revolutionized by the use of light for transmitting information. Scientists are already predicting that in the next decade many homes will have their own computers for shopping and banking, and an even more incredible possibility is seriously being studied: the use of the biochip — a tiny sliver of organic material — to replace the silicon chip in microelectronic communication devices.

To merely accept these developments in communications is to overlook the beauty of man's ability to turn the workings of nature to his own ends. To understand communications is to appreciate man's endless desire to explore his environment and tap its potentials for a better world.

The History of Communications

Long-distance communication began because of man's need for self-preservation, to signal the presence of food or warn of approaching danger. Primitive peoples' signals, bonfires, and drums, however, were limited in the amount of information they could pass on. A desire to communicate more complex messages spurred man on to use mirrors and flags to flash messages in code. But these could be used only at a distance where they could be seen. Finally, man began to use electricity.

Experiments with electricity took place well before people understood the scientific principles underlying it. In about 600 BC the Greek philosopher Thales discovered that when amber was rubbed it would attract lightweight objects. This was one of man's first scientific observations of static electricity.

It was not until the early nineteenth century that man combined his knowledge of electricity with a desire to communicate. In 1816 Sir Francis Ronalds built an experimental telegraph at his home in London. He placed metal plates at each end of wires 450 feet long and suspended balls near each plate. Each metal plate and ball represented a letter of the alphabet or a number. He electrically charged a glass rod by rubbing it with fur and then pointed it at a metal plate. The charge ran through the wire, attracted the ball, and signaled a letter or number at the other end.

Since the nineteenth century, electricity has been the communicator's chief tool. A major step central to communications came around 1800 when Alessandro Volta, an Italian scientist, invented his voltaic pile, an elementary kind of battery. The voltaic pile enabled Hans Christian Oersted, a Danish physicist, to discover in 1820 that electricity flowing through a wire could deflect a magnetic needle. A few years later, Joseph Henry, an American scientist, discovered that an artificial magnet could be made by passing electricity through a coil of wire wrapped around a soft iron core, now called the electromagnet. When the electricity was turned off, the core lost its ability to magnetize. Together these discoveries laid the foundations for the telegraph — the first practical means of communicating with electricity.

In 1835 Samuel Morse, an American scientist building on the discoveries of Volta, Oersted, and Henry, was able to transmit short bursts, or pulses, of electricity. Morse chose two pulses: a short pulse represented a dot, and a long pulse a dash. He then created a code by which these dots and dashes, used in various combinations, represented each letter of the alphabet and the numbers one through nine. In 1844 Morse transmitted a message between Baltimore and Washington, DC and demonstrated that messages could be sent electrically through wires.

A few years previously, in 1837, the British inventors William Cooke and Charles Wheatstone took out a patent for their five-needle telegraph. The surface of the instrument had the alphabet written over it and when switches were manipulated needles could be electrically deflected to spell out a message.

Another major development took place in 1876 when the American inventor Alexander Graham Bell patented his telephone. For the first time, Bell's telephone allowed people to speak with each other over long distances. But the telephone needed wires; what use were telephones and wires to people on ships?

In 1895 the Italian inventor Guglielmo Marconi successfully transmitted radio waves from one end of his house and received them at the other. The world took no notice of this important event, but in 1901 Marconi transmitted radio messages across the Atlantic Ocean from England to Canada, establishing that long-distance communication was here to stay.

Alexander Graham Bell opens the first long-distance telephone line in 1892. The line went from New York to Chicago, a distance of 900 miles.

Basics of Communications

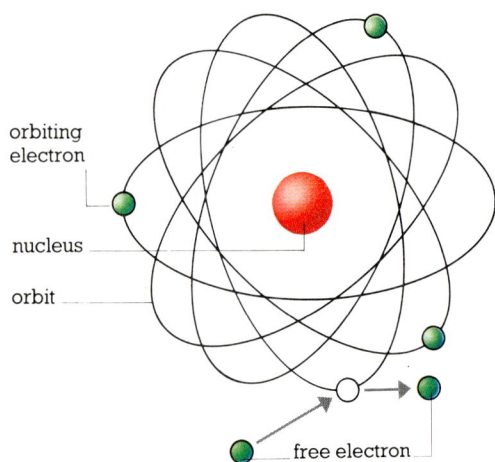

In this diagram of an atom, electrons orbit around a central nucleus. Sometimes an electron leaves its orbit (bottom) and becomes a free electron.

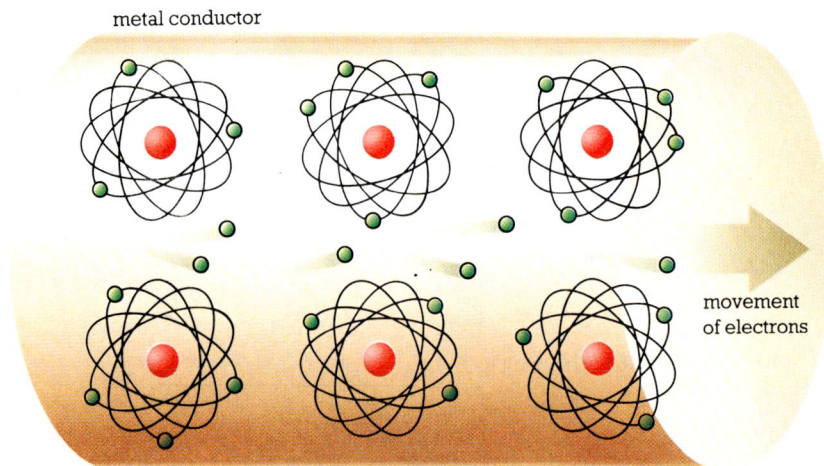

metal conductor

movement of electrons

In this diagram of electric current, voltage from a battery or generator causes a flow of free electrons from a negatively charged area to a positive one.

Electricity is the basis for all modern forms of communication, yet to most people it is only a mysterious force that conveniently lights lamps and spins motors. Scientists, however, know it is a fundamental form of energy that can be harnessed for human benefit in many kinds of matter. Consequently, to understand electricity it is first necessary to have an idea of the structure of the fundamental building block of all matter — the atom.

The structure of an atom can be loosely compared to the solar system. In the center of the atom, in place of the sun, is a tiny mass called a nucleus. In the nucleus are particles called protons, which carry a positive electric charge. Spinning around the nucleus, somewhat like planets, are even tinier particles called electrons. They carry a negative electric charge. The force that holds this tiny solar system together is the force of attraction between the two kinds of electric charge.

In the atoms of materials known as conductors, such as copper and aluminum, one or more of the electrons is very loosely held in its orbit, and is easily shaken free. Once free, an electron drifts randomly through the conductor until it is attracted by another atom that needs an electron. The random movement of free electrons, however, can be changed to a steady flow by applying a type of electric energy which is provided by batteries and generators. This flow of electrons is called electric current.

Electric current can be compared to the flow of water in the plumbing of a house. For the plumbing to work, there must be water in the pipes, the water must be under a certain amount of pressure, and follow a path from higher to lower pressure.

In an electric current, the free electrons in the conductor are like the water in the pipe, and the voltage is like the pressure on the water. The path is provided by an electric circuit; but instead of a path from high to low pressure, the circuit makes a path from negative charge to positive charge.

gamma rays

X-rays

ultraviolet radiation

visible light

infrared radiation

EHF

SHF

UHF

VHF

HF

MF

LF

radio frequencies

wavelength

amplitude

time

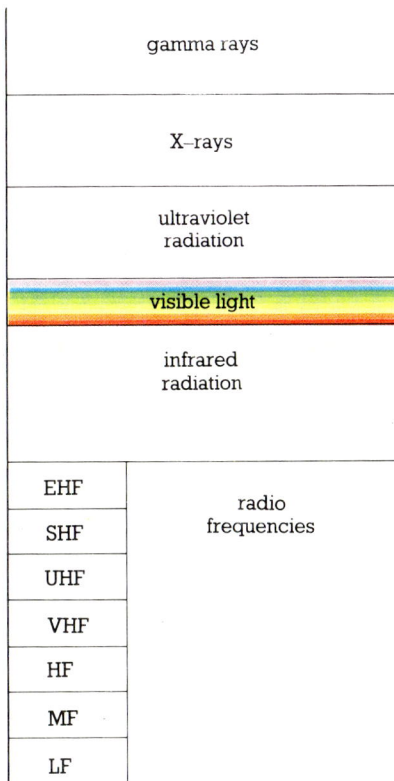

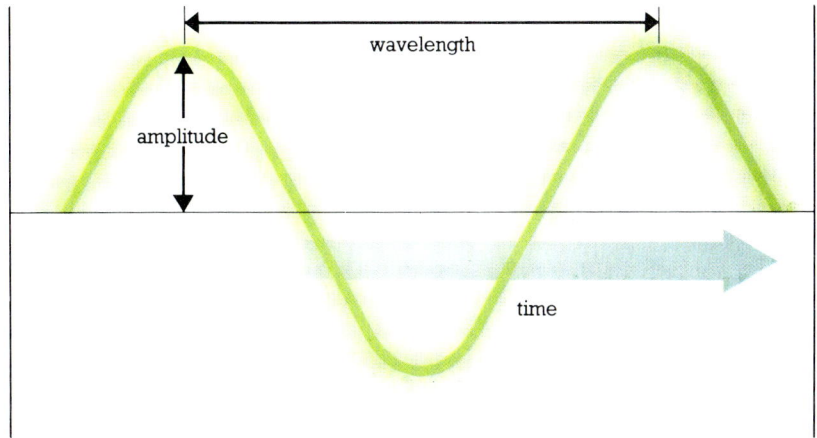

The electromagnetic spectrum includes all electromagnetic waves from very short to very long.

In an electromagnetic wave, the amplitude is the wave's strength, the wavelength the distance between two crests, and the frequency the number of waves per second.

Communications engineers use electric current to convey signals through conductors, most often copper wires. Telephones, telegraphs, and cable television, for example, all rely on electric current to carry information.

When electric current flows through a conductor the conductor is surrounded by a region of energy called an electromagnetic field. Ordinarily this field stays very close to the conductor. However, if the current is made to oscillate — that is, to change its direction of flow back and forth very rapidly — the field is cut off from the conductor, and the energy radiates outward in a fluctuating pattern much like a wave traveling through water. This is called an electromagnetic wave.

As electromagnetic waves move past a point, the energy level at that point fluctuates, forming crests and troughs. The difference between the crest or trough and the normal energy level is the wave's amplitude. The distance between the crests of two waves is called the wavelength; and the number of waves that pass a given point in one second is called the frequency.

Electromagnetic waves take many forms, including X-rays, visible light, and radio signals. When they are produced by oscillating currents, they form radio waves, which have relatively long wavelengths; shorter wavelengths, like X-rays and light, are produced by vibrating atoms or molecules, or by electrons changing orbits within an atom. Once generated, the only difference between electromagnetic waves is their frequency and wavelength.

Light and radio are two examples of electromagnetic waves used in communications. Radio waves have been used for decades to carry radio and television broadcasts. On the other hand, light wave communication, using hair-thin filaments called optical fibers, is quite new and may become the chief communications medium of the future — even over long distances, replacing electric current.

The Telephone

The telephone was one of the most important inventions of the nineteenth century. Although initially slow to come into widespread use, its progress since the turn of the century has been little short of astonishing. Over 500 million telephones are in use throughout the world today, and this number is increasing by over 100,000 every day.

Despite its many improvements, the type of telephone found in most homes today works in basically the same way as the device used by Alexander Graham Bell to make his famous call to his assistant Mr. Watson in 1876. In the mouthpiece of a telephone there is a microphone that contains granules of carbon — essentially the same material as the "lead" used in pencils. Carbon can conduct electricity, and compressing carbon granules has the effect of changing the strength of an electric current passing through them.

A person's voice is a combination of a large number of tones. In a telephone, sound waves generated by a person's voice exert pressure on a thin metal plate called a diaphragm, which vibrates and presses on the carbon granules. The changing pressure of the diaphragm on the carbon caused by the varying sound waves alters the strength of the electric current flowing through the carbon and creates an electrical pattern that matches the pattern of the sound waves. The fluctuating electric signals are then passed along wires to their destination. At the receiving end the electric signals flow into an electromagnetic coil in the earpiece of the telephone. The coil causes a second diaphragm to vibrate in harmony with the incoming signals and converts them back to sound waves, recreating the speaker's voice. The electric signals created at the mouthpiece and received at the earpiece are called analog signals because the strength of the signals varies in a continuous electrical "analogy" of the speaker's voice.

In the latest telephone systems another kind of signal, called digital, is used to carry conversations on the bulk of their journey through telephone wires. In a digital system, the analog signals coming from the telephone are converted into separate pulses representing the amplitude, or strength, of the original speech. A string of different pulses creates a complete word and, in turn, a conversation.

Engineers are continuously improving and updating the technology and performance of telephones. One such technological advancement is an electronic microphone that can be used in the telephone

In the earpiece of a handset (below) electric signals flow into a magnetic coil, that vibrates a metal diaphragm to produce sound. In the mouthpiece, a diaphragm vibrates with the sound waves, pressing on a cylinder filled with carbon granules, changing the current to match the speaker's voice.

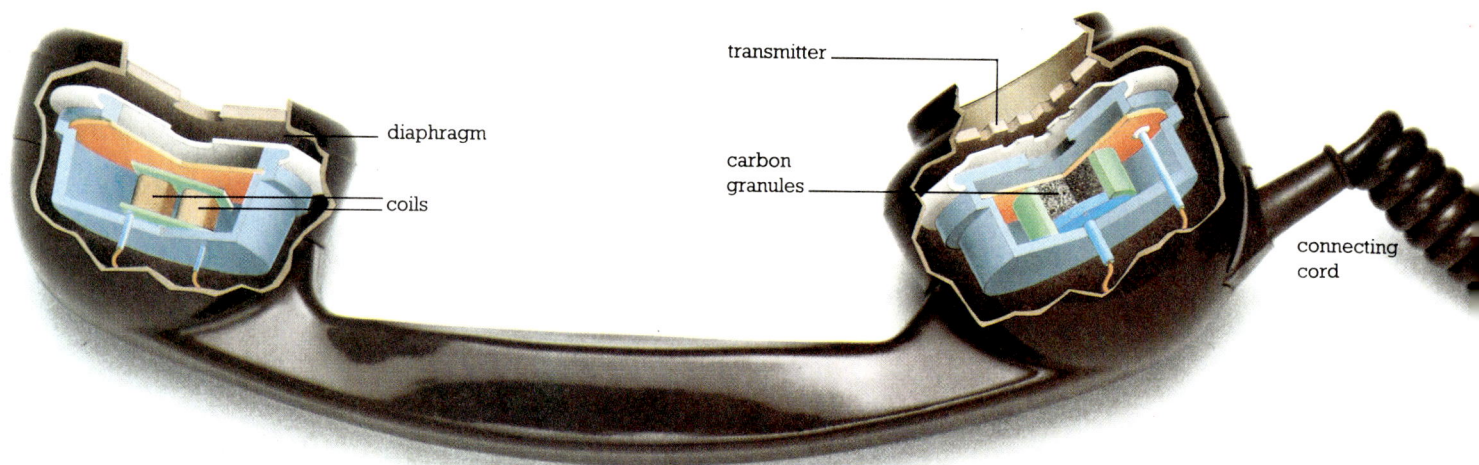

diaphragm

coils

transmitter

carbon granules

connecting cord

mouthpiece which eliminates the need for carbon granules and gives much better voice quality than the traditional version. Another development soon to be in widespread use is the cordless telephone. This usually has a small radio transmitter in the handpiece in place of the wire that connects it to the body of the instrument. The radio transmitter sends dialing and voice signals to a base station that is connected to the telephone lines. The advantages of a cordless telephone are many. For example, a person can carry the telephone to different rooms of his house or even outside of the house, and still make and receive calls.

New and exciting uses of the telephone are already available. For example, the development of digital signaling has made it possible to verify a customer's credit card by passing the card through a special reader attached to a telephone. Electronic circuits then read the card's magnetic code and dial the computer center of the company concerned. Once the status of the cardholder is known, it is displayed on a screen seen only by the person checking the card. The entire operation takes place in seconds.

There are now telephones that show the time, can be used as calculators, and store numbers to which calls can be diverted when the owner is not in. Picture phones on which callers are seen as well as heard are now technically possible. Alexander Graham Bell would scarcely recognize his invention.

terminals

capacitor

dial

bell

bell

The quaint device above is the Ericsson Magneto tabletop telephone, introduced in 1892. Turning the crank at left rang a bell at the switchboard, alerting an operator that the user wished to place a call.

The body of the telephone is a fairly simple mechanical device. Inside are wires, switches, circuits and bells. Some of the wires provide power to the handpiece, bell, and dial. Others allow signals to be received and transmitted.

11

Telephone Switching: I

The telephone is the most efficient device man has ever developed for communicating over long distances. This efficiency depends on how quickly one person can make contact with another. The machines responsible are the telephone switching systems.

The simplest telephone switching systems were merely large boxes or switchboards, that housed wires, jacks, and plugs from different telephones. To

A crossbar system receives telephone signals and holds them until the complete number is received. A path is then made through the switches to the person being called.

make a connection, an operator plugged the cable of the person making a call into the appropriate set of telephone lines. Such systems were efficient as long as operators did not have to cope with many calls.

As more telephones came into use and more people needed fast connections, the entire process was mechanized and became known as an automatic telephone switching system. The first steps to produce such a system were taken in 1889 by Almon B. Strowger, an American undertaker. It is said that Strowger believed his competitors were bribing telephone operators at the local switchboard to divert his calls, and he responded by inventing an automatic switch made from a shirt collar and hat pins.

In its modern form, the Strowger switch, commonly called a step-by-step switch, consists of a semicylindrical bank of electrical contacts arranged in a grid. A selector arm attached to the switch bank moves up or down. As a person dials a number, the movement of the dial sends electric pulses through the telephone wires to the switching center. Each user's telephone is connected to a step-by-step switch in the switching center. The incoming electric pulses — or the numbers dialed — make the selector arm move to different contacts on the bank of switches. Once the dialing operation is completed an electrical conducting path is formed through switches in the system. An electric current flows along this path, ringing the telephone bell at the receiver's end. When the telephone is lifted, the lines between each handset are connected.

A major development in telephone switching took place in 1938 when crossbar switching was first used in the United States. With crossbar switching, two sets of electromagnets establish a path through a series of overlapping contacts. In a step-by-step system, each digit makes a new contact immediately after it is dialed — in a crossbar system as each digit is dialed it is held in storage. Only after the complete number has been dialed and received is the final connection made to the person being called.

Step-by-step and crossbar systems are electromechanical, which simply means that an electric impulse triggers a mechanical switch. In the 1950s telephone systems engineers began to experiment with the possibility of electronic switching systems. It seemed that electronic switching might not only switch calls faster, but also provide new services.

Although telephone engineers knew what their goal was, it was not until the late 1950s that an electronic switching system seemed possible, and even then there were several false starts. Switching centers were built in the laboratory, using vacuum tubes as switches. At the time, tubes were being used successfully in radio and television receivers, but in telephone systems their effectiveness was limited. A switching system having hundreds of switches and vacuum tubes needs a great deal of power to drive them. These first experimental switching centers consumed huge amounts of electric power and generated a great deal of heat. They were clearly not practical, but they gave engineers a great deal of useful knowledge.

The solution to the fully electronic switching system finally appeared in 1948 when a team of scientists in the United States invented the transistor. This development ultimately led to the creation of the microchip and heralded the modern telephone switching system of today. Now computers not only switch but store telephone numbers, make automatic calls and interrupt a speaker in an emergency. From a shirt collar and a few hat pins, the telephone system has entered the age of high technology.

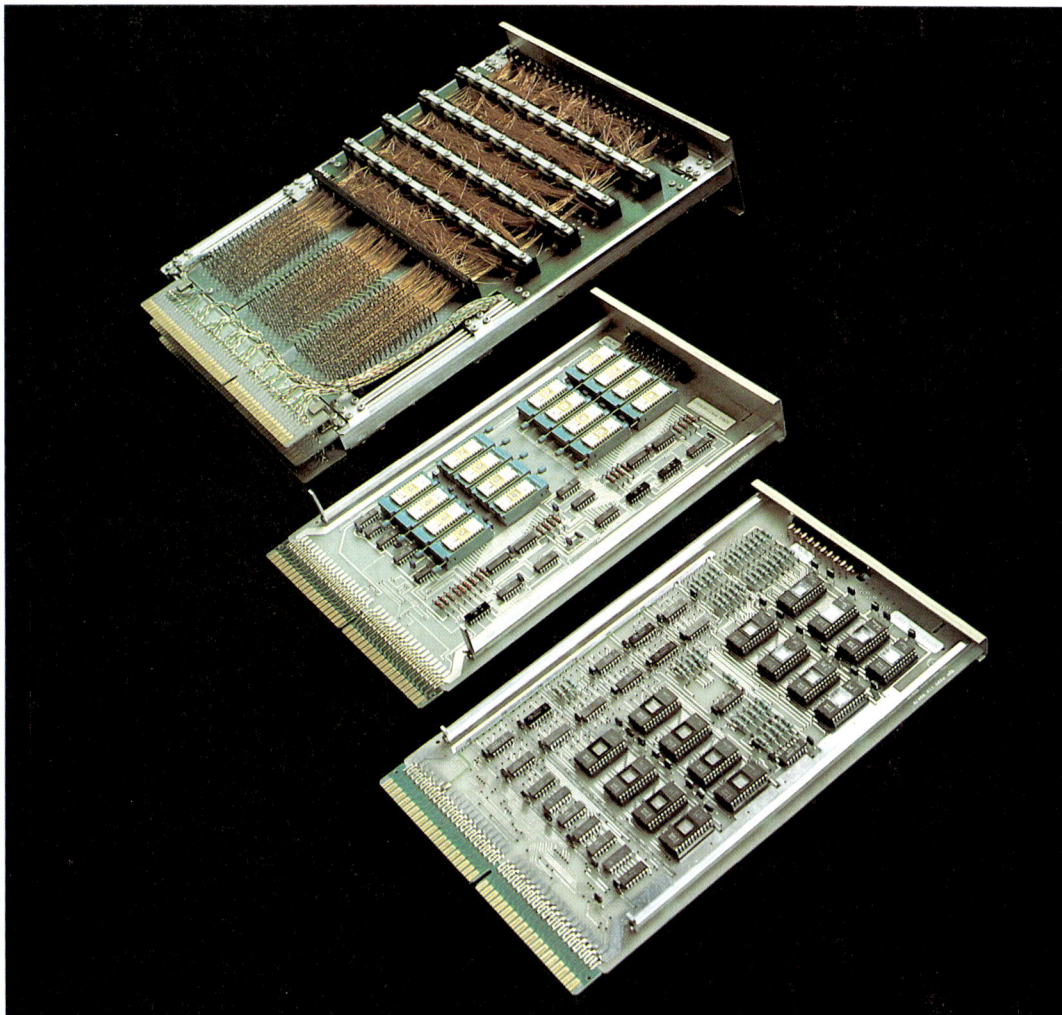

Strowger switch

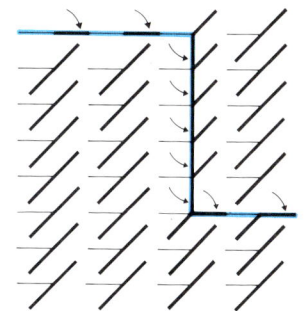

crossbar switch

A step-by-step switch (top) and a crossbar switch (bottom).

Early hand-wired switching circuitry (left, at top) with two generations of integrated circuitry.

Telephone Switching: II

Today's telephone switching systems are actually specialized computers. Most are automatic and have no moving parts. The modern telephone system grew from the transistor — the first important solid-state electronics device — into an area of technology now known as microelectronics.

Telephone signals are analog signals, or continuously varying electric waveforms. In the 1960s, digital signals — pulses of electricity that represent the information of the original speech — began to make their appearance in the United States. A digital signaling system has several advantages over an analog system. With a digital system, a telephone line can handle many more calls simultaneously and with less interference than with an analog system.

The most common form of digital coding is called pulse code modulation, or PCM. With PCM voice signals are converted into digital form by splitting each telephone conversation into millions of pulses. Groups of pulses from a conversation are carried on the telephone wires between switching centers. Multiplexers mix the pulses of up to twenty-four conversations into one signal that is then sent down a line.

When a group of conversation pulses arrives at a digital switch, they are stored for a fraction of a second, sorted out into groups with the same general area address by devices called demultiplexers, and sent on their way. This occurs at each switching center, until the conversation reaches its destination.

This is a complicated process even when the conversations travel a direct route between two centers. In practice, most telephone calls pass through many more than two switching centers. In some countries these are old-fashioned electromechanical analog switching systems that pose a problem because the digital signals have to be converted into analog form, switched through the center, and reconverted to digital pulses for transmission to the next switching center. In the near future most telephone systems will

On the oscilloscope screen below, the discontinuous signals of a telephone conversation in pulse code modulation, or PCM, appear as a stream of intermittent rectangles.

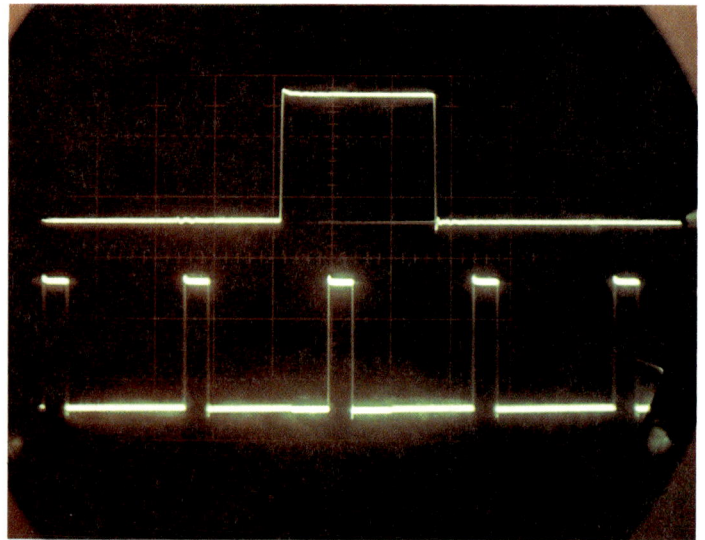

be able to handle digital signaling and the advantages of this breakthrough will be fully realized.

Some older electromechanical systems now use computers. Modern electronic switching systems are actually computers programmed to do routine useful chores. One type of computer system is an Electronic Switching System, or ESS. With ESS the instructions for operating the system are held in a computer memory. Pulses dialed from one telephone are collected and analyzed by the computer which then orders switches to make the connection.

The chief advantages of ESS systems are that engineers can change the operating instructions of a center without altering the switching equipment. ESS also gives subscribers many options such as automatic dialing and repeat calling. With the latter a system will keep dialing a busy number until it is free and

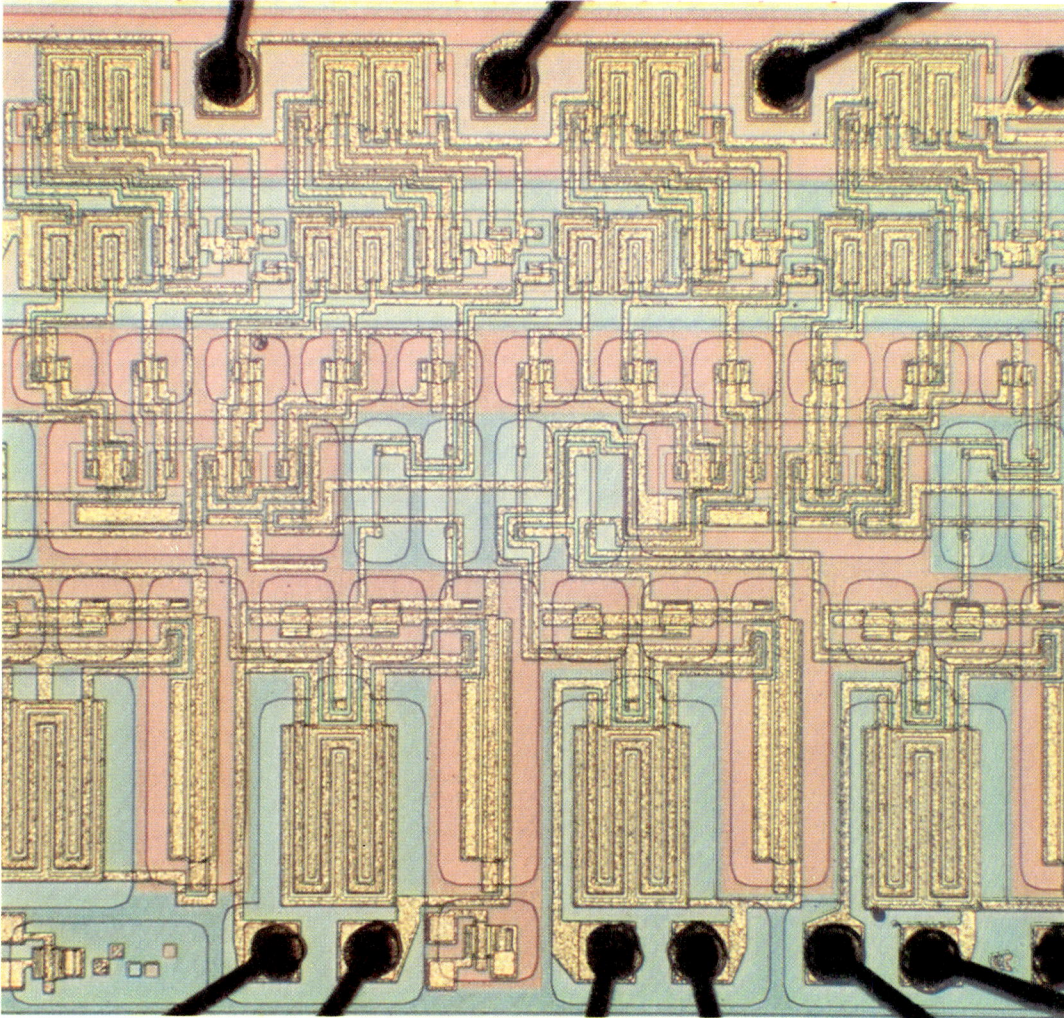

On this microelectronic circuit, thousands of transistors are interconnected on a piece of silicon one-quarter inch square.

In the picture below a technician monitors a telephone switching center where phone calls are directed to their destinations. A central control unit, a kind of computer memory, handles the flow of telephone traffic.

then connect it to the telephone initiating the call. These ESS services are not yet in widespread use.

Until the 1960s, telephone systems used electromechanical switches and could handle only analog signals. Today electronic switching systems are in wide use in the United States and can switch both analog and digital signals. These systems rely on solid-state switches. This means that instead of a mechanical contact creating a path through a center as in the step-by-step system, solid-state switches create the path of the signal through banks of circuits.

Why have engineers designed systems that seem more complicated than the old style systems? ESS switches provide considerably better telephone connections than electromechanical machines, yet occupy only about fifteen per cent of the space. They are also less expensive for the customer.

The Telephone System

A telephone is useful only if it is connected to other telephones, and so vast, complex networks are needed to allow people to communicate with one another. A local switching center is used to link a caller to a long-distance switching center. If a call is made over a short distance it may simply be sent from one local center to another. Over greater distances the call will pass through a local switching center into the long-distance center, then through the network to the local center of the number dialed.

Telephone networks vary from country to country. There is, however, one thing all networks have in common: each telephone has a unique number. In the United States, that number has ten digits. The first part is the area code, which corresponds to a particular region, of which there are 152 in the North American telephone system. The next three digits identify a local switching center in that area. The last four digits direct a call to an individual telephone.

In the United States the system known as DDD (Direct Distance Dialing) allows a customer to dial any place in the country without the assistance of an operator. Dialing the area code alerts the switches in the local switching center to select a line to a toll (long distance) switching center. A line then connects this local toll office with one in the area being called. The next three numbers tell the distant toll office to open a line to the local switching office in that region. The last four digits direct a call to its final destination.

When making an international telephone call, a special set of code numbers is dialed before the other parts of the number. The local switching center interprets these special numbers and routes them through a toll office to an international center.

The main difference between a local and a long-distance center is one of capacity. Long-distance switching centers are connected to one another and to local centers by multiplexed lines that can cope with many thousands of telephone calls at once. In countries with older telephone networks, local and long-distance switching centers may be slow because they still use mechanical switches.

The telephone lines that link systems range from wires slung on poles to specially protected cables buried in ducts underground. Cables absorb some of the energy of the telephone signal as it passes down a line, and so all cables need devices called repeaters used to amplify and boost signals at regular intervals.

Most international calls are carried by huge cables laid on the sea floor. The construction of an undersea cable capable of crossing an ocean calls for high precision engineering and sophisticated manufacturing techniques. Undersea coaxial cables must be waterproof and armored to prevent accidental damage. They also need repeaters every six miles or so.

Undersea cables offer high-quality communication but communications satellites offer other advantages, such as eliminating the massive number of repeaters usually required with cables. One satellite can connect subscribers thousands of miles apart, cutting costs when oceans have to be crossed.

Telephone networks are beginning to replace the copper cables used today with cables made of very pure glass. In order to travel through these so-called optical fibers, electric signals are converted into pulses of light and then bounced through the glass fibers to their destination, where they are converted back into electric signals.

During the journey of a long-distance call, the electric impulse that carries the voice is directed through several switching centers to reach its destination.

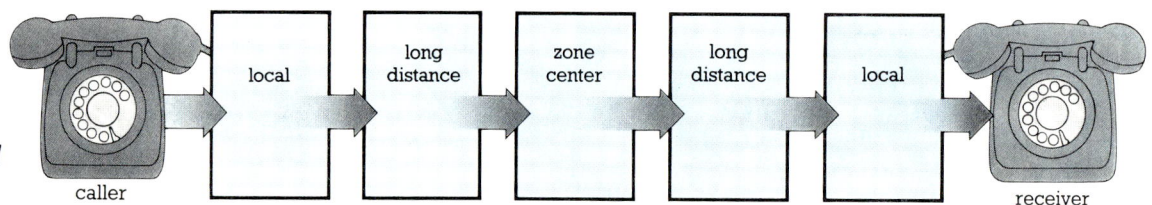

caller | local | long distance | zone center | long distance | local | receiver

The jungle of wires below is only one small part of an electromechanical switching system.

A line of telephone poles (left) troops across the countryside bearing cable between relatively distant destinations. The number of cables needed in cities makes poles unsightly and impractical, and so cables are often run underground.

Cables

fiber optic cable

copper cable

coaxial cable

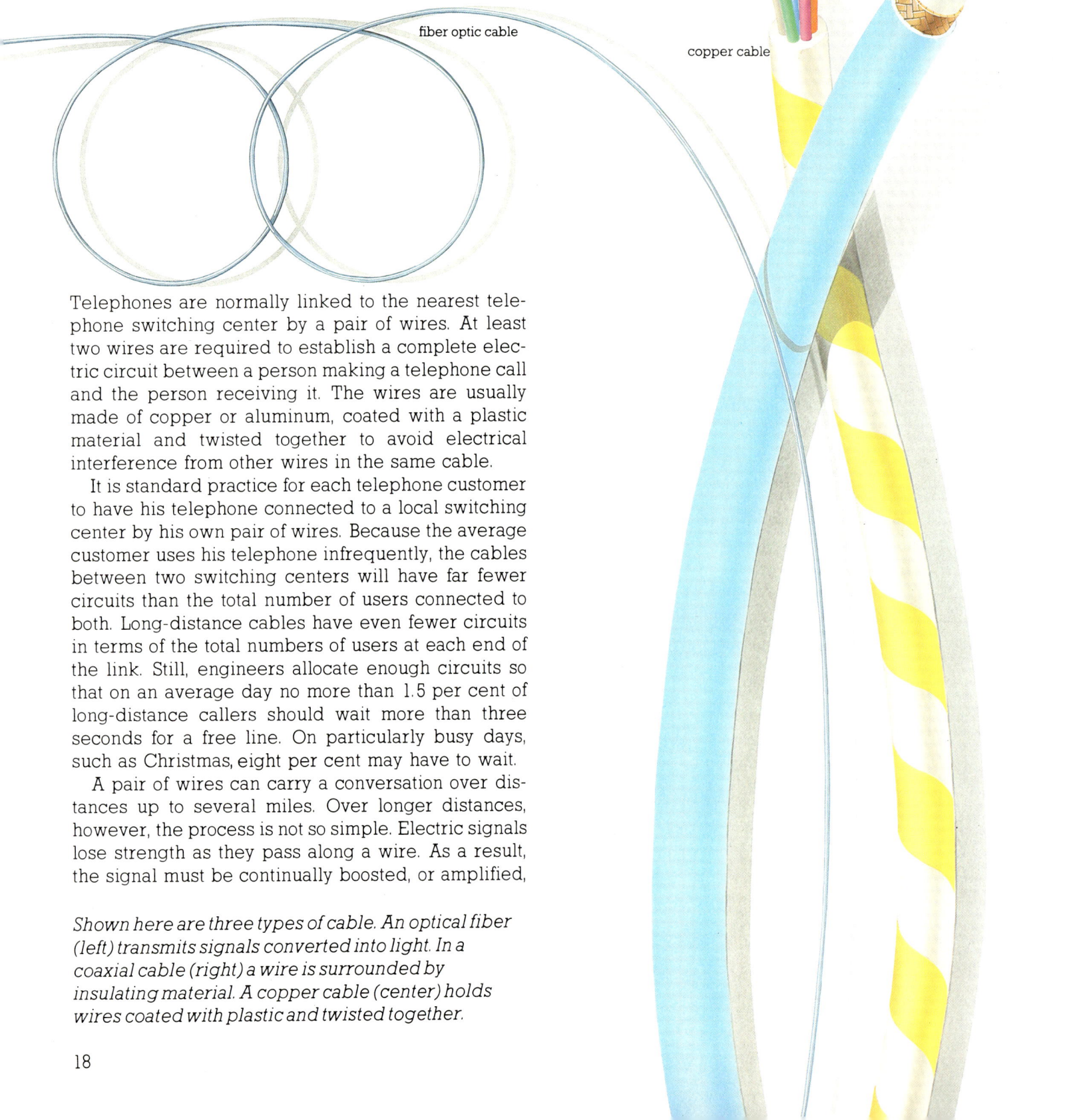

Telephones are normally linked to the nearest telephone switching center by a pair of wires. At least two wires are required to establish a complete electric circuit between a person making a telephone call and the person receiving it. The wires are usually made of copper or aluminum, coated with a plastic material and twisted together to avoid electrical interference from other wires in the same cable.

It is standard practice for each telephone customer to have his telephone connected to a local switching center by his own pair of wires. Because the average customer uses his telephone infrequently, the cables between two switching centers will have far fewer circuits than the total number of users connected to both. Long-distance cables have even fewer circuits in terms of the total numbers of users at each end of the link. Still, engineers allocate enough circuits so that on an average day no more than 1.5 per cent of long-distance callers should wait more than three seconds for a free line. On particularly busy days, such as Christmas, eight per cent may have to wait.

A pair of wires can carry a conversation over distances up to several miles. Over longer distances, however, the process is not so simple. Electric signals lose strength as they pass along a wire. As a result, the signal must be continually boosted, or amplified,

Shown here are three types of cable. An optical fiber (left) transmits signals converted into light. In a coaxial cable (right) a wire is surrounded by insulating material. A copper cable (center) holds wires coated with plastic and twisted together.

every few miles by devices called repeaters. The first use of repeaters took place in 1915, when a telephone signal was transmitted across the United States by using large vacuum tube amplifiers.

Another important achievement in long distance transmission was the development of the coaxial. A coaxial consists of an outer conducting tube, about three-eighths of an inch in diameter, with a conducting wire in its center, held in place by disks of plastic for insulation. The coaxial gets its name from the fact that the wire is at the very center, or axis, of the tube. Up to twenty coaxials can be bunched into a coaxial cable, which is then sheathed in plastic and lead for protection. The construction of coaxial makes it suitable for transmission of high-frequency electric waves, so that circuits carried over coaxial cables can use frequency division multiplexing, a technique that permits thousands of conversations to be carried over each circuit simultaneously.

Frequency division multiplexing works by using the low-frequency electric wave created by the speaker's voice to modulate a high-frequency carrier wave. If the cable can carry a wide enough range, or band, of frequencies, then it can carry many different conversations simultaneously, each assigned to a narrower channel of frequencies within that band. A typical coaxial cable consisting of ten paired coaxials can provide as many as 3600 voice channels per circuit, or 32,400 conversations per cable, with one pair kept as a spare.

Coaxial cables are also used for undersea telephone links. These cables are armored to prevent damage from ships in shallow areas near land. Highly reliable repeaters powered from land are installed in the cable at about six mile intervals.

Newly developed fiber optics, however, may not only surpass conventional cables but may revolutionize all types of present day communication services and set the scene for a range of new ones. Video telephony, high-speed facsimile, supertelex, and a variety of other digital services can be carried on fiber optic cables that are much thinner than their metal equivalents but nevertheless have a much larger traffic-carrying capacity. In a fiber optics system the number of repeater stations necessary to boost signal strength is far fewer than with metal cables. In the near future it is likely that all metal cables will be replaced by fiber optic cables.

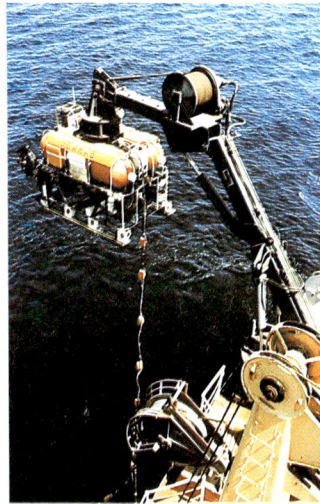

A special submersible craft lifts an undersea cable from the ocean floor in order to service a repeater. Although built to last twenty-five years or more, these cables still need maintenance.

In one of the first trials of undersea fiber optics, a specially equipped ship lays cables on the floor of a Scottish loch. The results have been encouraging, and engineers hope to lay the first transatlantic fiber optic cable by 1985.

Facsimile Systems

Facsimile is the process of transmitting words and pictures by wire or radio. This form of sending information existed even before the invention of the first telephone. Thirty years before Alexander Graham Bell demonstrated his remarkable creation, inventors had devised a form of facsimile communication by using levers to mechanically transfer a picture or image, but only over a few miles.

Modern facsimile machines work by electronically scanning a document or photograph. The scanning head of the machine moves from side to side down the page to be transmitted. As it moves, the head illuminates the document. Inside the head, sensors measure the light reflected off the illuminated page, sensing tones of white, off-white, gray and black.

Facsimile machines of the so-called analog variety convert these measurements of reflected light into electric signals. The signals, which represent electronically what is on the page of the original document, are then sent over a conventional telephone line. At the receiving end the signals activate a printing head that moves across and down a sheet of paper, creating a facsimile — or exact copy — of the original. Digital facsimile machines operate in much the same way as analog ones but the signal is a code of electronic pulses rather than an electrical interpretation of the tones of black and white on the page.

For years newspapers have used facsimile to transmit photographs. Increasingly, publishers of newspapers are using special facsimile machines and communications satellites to transmit text and pictures to printing factories hundreds of miles from their offices. Using facsimile in this way is not only faster but cheaper than shipping newspapers by road, rail, or air, and allows publishers to produce international editions in which local material can be mixed in with that written in the main office.

The ability to transmit information quickly is only one reason for using facsimile. It can be extremely difficult, for example, to send messages in foreign languages by other types of communications systems. Using facsimile saves time and eliminates the inevitable mistakes introduced by telex or telegraph operators. Telex, for example, has approximately fifty keys for letters, punctuation and numbers, but some languages, such as Chinese and Japanese, have about 50,000 characters.

Facsimile is used to send updated weather and navigational maps and charts to ships at sea. In some countries a facsimile copy of a legal document can be used in place of the original. A facsimile copy of a will, for example, can be immediately acted upon in the same way as the original.

Despite its many advantages, facsimile has so far

In this diagram of a facsimile machine, light is focused by lenses onto an image on a drum. The light reflects and is bent by a mirror through an aperture. A light-sensitive diode changes the light into electric signals that are passed to the receiving end.

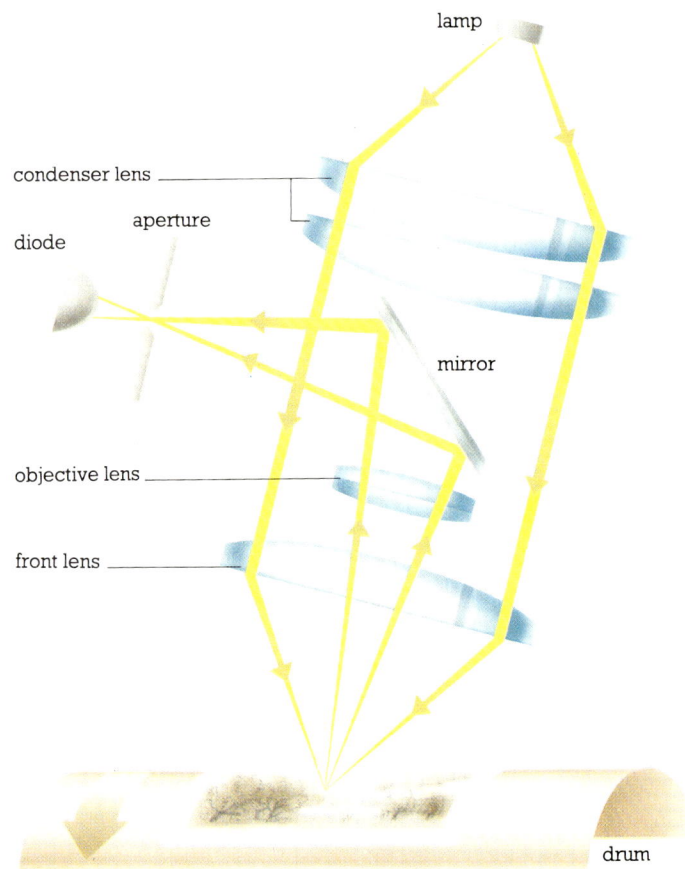

failed to make an impact on its biggest potential markets — business and domestic communications. Initially the reason for this failure was because early facsimile machines were slow and the different makes of machines were unable to communicate with each other. Engineers and administrators have since eliminated such problems.

The most remarkable improvement in facsimile has been the speed with which documents can now be transmitted. Today's facsimile machines can transmit all the information contained on these two pages to the other side of the world in about one minute. Some machines can transmit a page every fifteen seconds, and the most advanced can transmit in color — although this might take a few seconds longer to get from one place to the other.

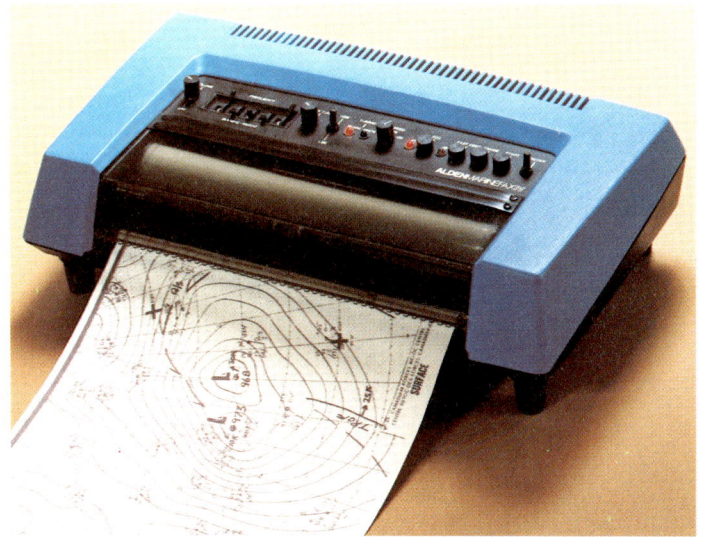

The small facsimile machine above can produce an accurate replica of a weather map transmitted from a distant weather station. The map informs pilots and sailors about the precise weather conditions at their destination in advance of their arrival.

The facsimile machine in this news office transmits and receives pages of pictures and text to and from outside offices. News organizations often rely on facsimile to obtain pictures quickly from all over the world.

Telex and Supertelex

This 1858 telegraph had a tape-punching device, transmitter and pen-type receiver. Telegraph messages coded in advance on punched tape could be sent much faster than those tapped out in Morse code by a human operator.

A journalist rushing to file a news story or a businessman anxious to close a deal is more likely to head for the nearest telex machine than to pick up the telephone. Telex is a fast, efficient means of transmitting written words and numbers. When sending and receiving, the message is seen rather than heard so there is less chance of misinterpretation or misinformation being passed on.

Telex is the descendant of the telegraphy systems devised by Samuel Morse and Charles Wheatstone. These early forms of telegraphy provided a means of sending a coded message by an electric current over a wire. At its destination, the receiver used the code to reconstruct the message.

A telex machine mechanizes the coding, transference, and decoding of a message. To send a message the operator types out words and numbers on a special keyboard. As the message is being typed, a series of small holes based on a special code are punched into a thin paper tape. The tape is then run through a device known as a reader that converts the code into electric signals which are then sent down an ordinary telephone line to their destination. At the receiver's end the process is reversed and an automatic typewriter prints out the message as it was originally typed. If an immediate answer is needed, the message can be sent as it is being typed and the receiving operator can type an immediate reply. This process is often used for hotel and airline reservations for fast travel arrangements.

Telex messages are automatically coded and decoded conforming to an international standard — used in most countries — known as the International Telegraph Alphabet, or ITA2. ITA2 codes all of the letters and numbers on a telex keyboard into combinations of five positive or negative electrical pulses. The letter A, for example, is coded as two negative and three positive pulses; B as one negative, two positive and two negative pulses, and so on. Approximately 800,000 telex machines around the

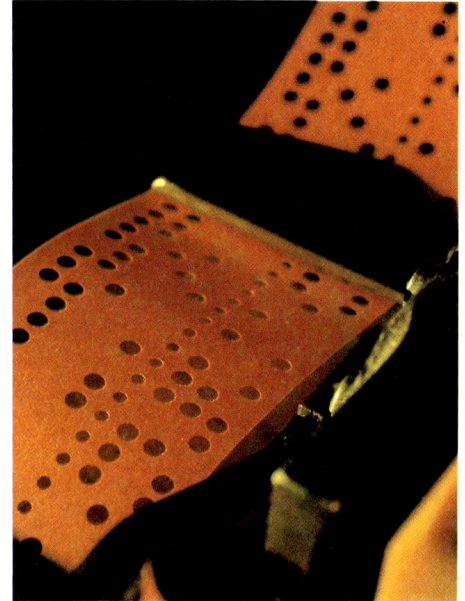

world chatter in ITA2. While it is possible to send messages on special telex machines without using ITA2, this is a slow and expensive process.

In some modern telexes tiny computers inside the machine can automatically check to see if the receiving telex is ready to receive a message and will let the operator correct or change the message before it is sent. Many such technological advances have led to the creation of what is known as supertelex or teletex. Today, a supertelex service has already been started in Britain, Germany, and Sweden. By 1984 most European countries, Canada, Australia, and the United States will have such services. Supertelex is as versatile as a word processor or an electronic typewriter and faster than the traditional telex, transmitting up to 3000 characters in ten seconds instead of the three and one-half minutes required by the standard machine.

Supertelex has a keyboard, display screen, and a fast, high-quality printer. An electronic controller can do many of the jobs previously done by a human operator. While being typed by an operator, the message automatically appears on a display screen. The operator can then alter the text by deleting characters or words, moving paragraphs around, leaving spaces for pictures, and so on. On some machines the operator loads the message into a computer memory by pushing a special key. The machine may then be told by the operator where to send the message, and an electronic controller dials the necessary codes and instructions. The message is then sent to a similar computer memory at its destination where it is decoded and printed.

Engineers believe that the supertelex of the future will be even more versatile. Soon you will be able to use supertelex to talk to computer terminals and send material to facsimile machines. But the supertelex of the future will not be unfriendly — it will still be able to talk to its old-fashioned colleague, the traditional telex machine.

Radio: I

In the early nineteenth century the Danish scientist Hans Christian Oersted noted that magnetism was created around a wire carrying an electric current. Michael Faraday, an English scientist, soon discovered that the reverse was also true: moving a magnet past a wire would create a current in the wire.

In the second half of the century the Scottish physicist James Maxwell mathematically described the phenomenon of electromagnetism, as it became known, and theorized that it could travel through space as waves. Later, the German scientist Heinrich Hertz proved these theories experimentally.

In the 1890s the Italian inventor Guglielmo Marconi succeeded in using electromagnetism to serve communications. In England in 1896 he was able to send signals over a distance of nine miles. In that same year he was granted the world's first patent for a wireless telegraph system.

At the beginning of the twentieth century radio was primitive and inefficient. To transmit radio waves an electric spark was created which generated powerful oscillating electric currents in an antenna. Part of the electromagnetic energy surrounding the current then radiated out as electromagnetic radio waves. Receiving equipment was equally primitive. It had been discovered that loose iron filings would stick together, or cohere, when an electric spark was discharged nearby. By connecting the coherer — a small glass tube filled with iron filings — to an antenna, it was possible to detect a spark at long range. However, the filings had to be shaken apart before any new transmission could be received.

Despite the limited capabilities of his early equipment, Marconi proved that radio waves were a practical alternative to wires for carrying telegraph signals over short distances. On December 24, 1906, the American scientist Reginald Fessenden used his radio equipment to broadcast music and speech to ships stationed along the coast of Massachusetts.

There are several ways to transmit sounds using radio waves. The simplest and most widely used technique is to vary the strength of the radio wave with the fluctuations in the sound waves picked up by a microphone. This process is known as amplitude

Guglielmo Marconi demonstrates his radio transmitter in Chelmsford, England, in 1919. Bulky and low-powered, it nevertheless heralded the birth of modern radio broadcasting.

The transmission of an AM radio signal is shown below. A microphone changes sound into electric waves to modulate a high-frequency current. The signal is amplified, passed to an antenna and sent out as radio waves.

In the receiver, a tuner matches the frequency of the transmitted signal and a detector separates the audio part of the wave. The detector current is amplified and passed to a loudspeaker converting the wave into sound.

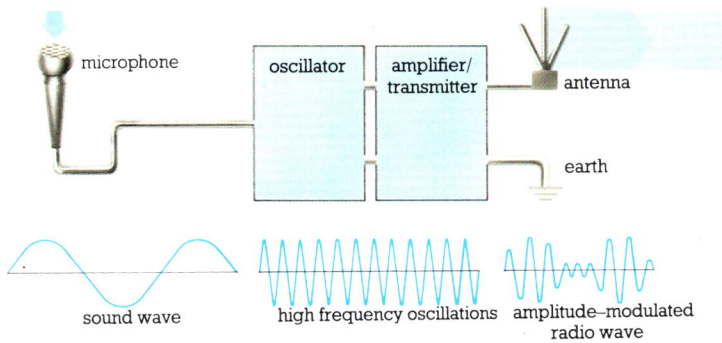

microphone — oscillator — amplifier/transmitter — antenna — earth

sound wave — high frequency oscillations — amplitude–modulated radio wave

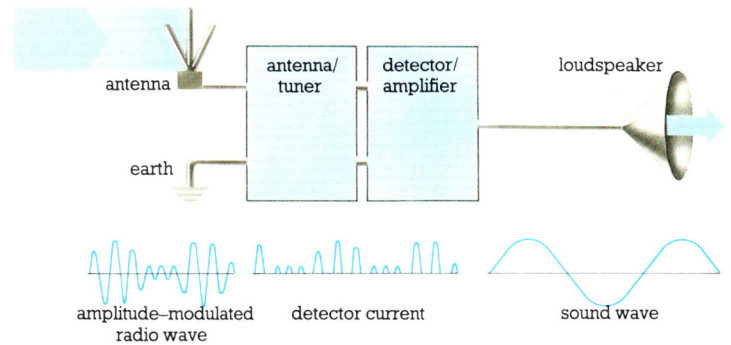

antenna — earth — antenna/tuner — detector/amplifier — loudspeaker

amplitude–modulated radio wave — detector current — sound wave

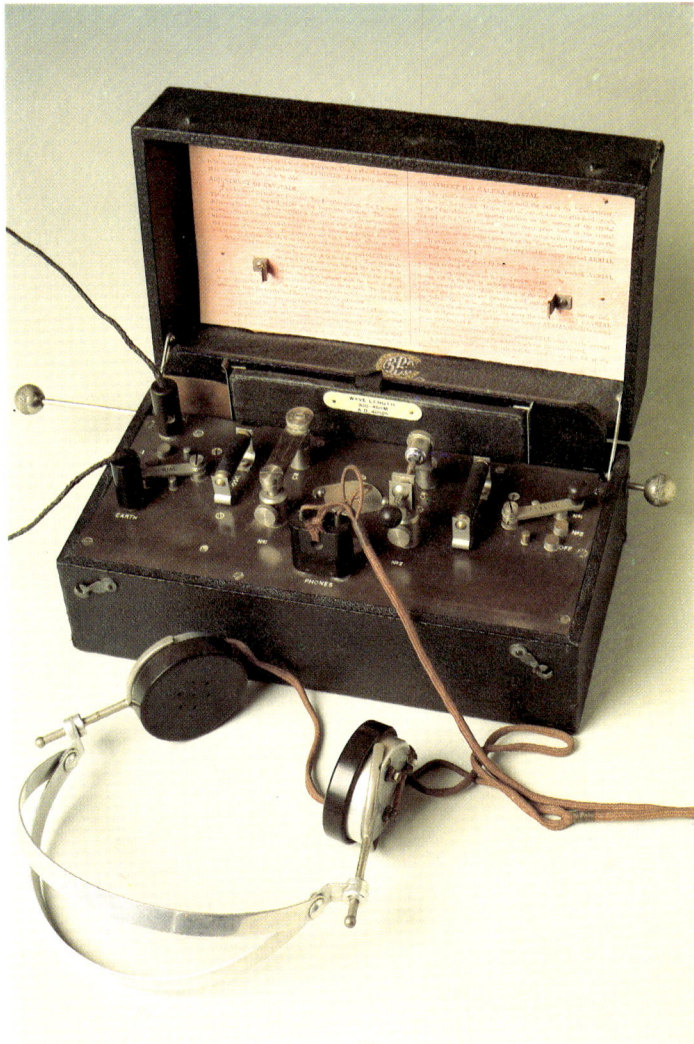

modulation, or AM. At the receiving end the process is reversed and the electromagnetic waves are changed back into sound waves. Using this method, in 1920 popular commercial broadcasting began. No longer was radio only in the hands of the scientist — it was now heard in the home.

Early broadcast receivers used crystals to detect incoming radio signals. As manufacturers learned how to mass-produce vacuum tubes, these so-called crystal sets gave way to radios capable of being tuned to a wide range of frequencies. By 1936 the vacuum tube had paved the way for television.

In their search for better electronics equipment components, a small team of engineers in the United States began working in the late 1930s toward a device that was to produce one of the biggest revolutions in twentieth century technology: the transistor. The first transistor, invented in 1948, was made from germanium, an element obtained during the refinement of copper, lead, and zinc ores. Later, scientists and engineers developed transistors using silicon, an inexpensive material that can be made from sand. The transistor could do almost everything that a vacuum tube could at a fraction of the cost. Furthermore, it does not generate masses of heat — an expensive and inconvenient by-product of the vacuum tube.

Long before transistors, people relied on crystal sets such as the one shown here. These had a simple tuner, detector, and no amplifier. A huge antenna and earphones were needed to hear signals.

Radio: II

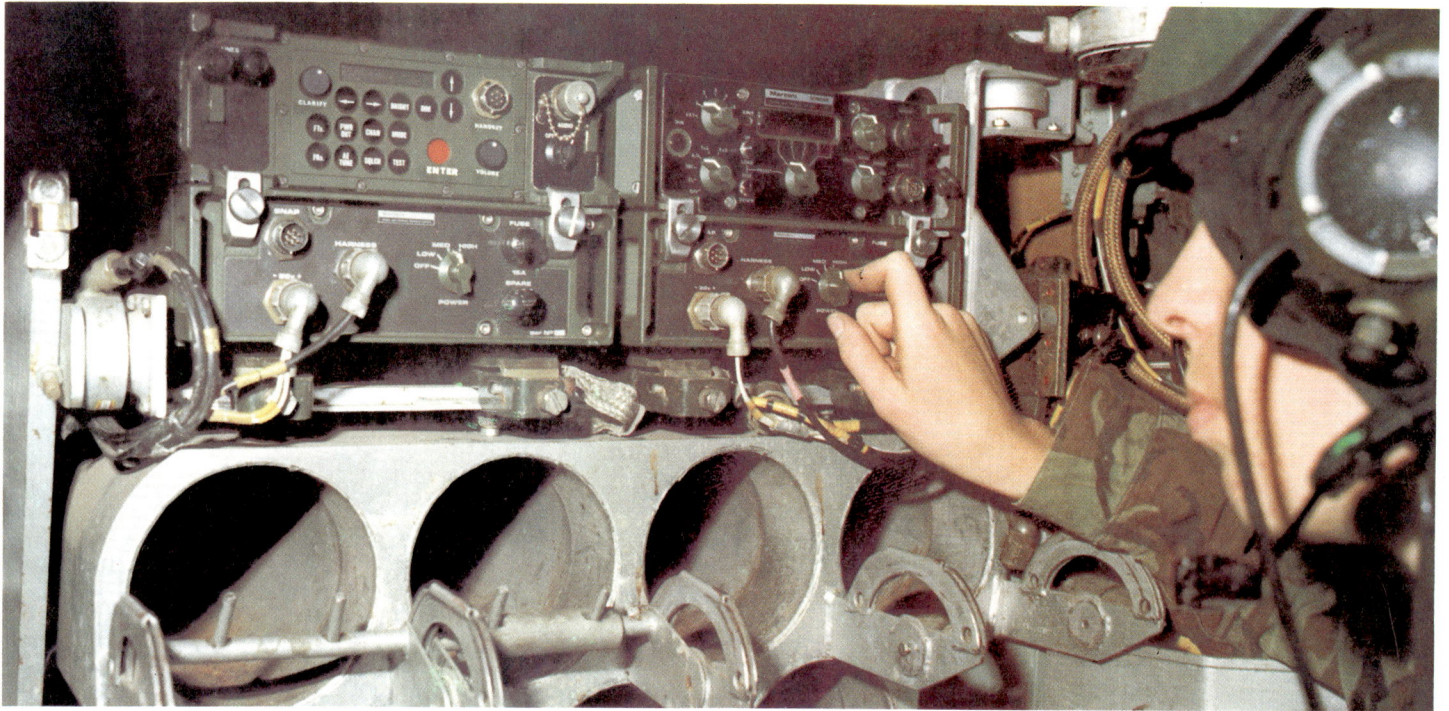

When the transistor was announced in 1948 few people could have foreseen the impact that it was going to have on their lives. Although transistor, or solid-state technology as it is now known, had a slow start, the last decade has witnessed dramatic advances in the capabilities of all manner of electronics equipment — and radio communications equipment is no exception to this.

It is now technologically possible for private citizens to talk to one another all over the world by satellite — although this requires expensive equipment. Soon it may be possible for people everywhere —and not only radio enthusiasts— to communicate by radio rather than by writing letters or using the telephone. First, however, manufacturers must produce inexpensive, two-way radios that the general public can afford.

That time may not be far away. Manufacturers are already investigating the possibility of producing and selling wristwatch radio transmitters and receivers that can communicate across continents and oceans by bouncing signals off satellites. They are also looking toward mobile radio computer terminals no larger than a book, and research teams are developing silicon chips that can even recognize and respond to the human voice.

Before the transistor, radios, transmitters and receivers were large and cumbersome. A typical community broadcasting transmitter was ten times the size of today's equivalent — roughly the size of an office filing cabinet. Most receivers, except for military use, were bulky and certainly not portable.

Early transistors were the size of a vitamin capsule — fifty times smaller than the first vacuum tube. But for a circuit with hundreds of transistors, even this is too large. Engineers now can put hundreds of thousands of transistors onto a microchip, a sliver of silicon one-quarter inch square.

At the site of an emergency, a policeman (below) communicates vital information to his headquarters with a mobile radio. The information can then be routed to other services, such as firemen.

In a commercial radio station (above) a disc jockey reads from a script. The many levers covering the console in front of him are used to relay instructions.

As engineers are able to fit more and more transistors onto a single microchip, more computing power can be built into communications equipment. Some microchip radios, for example, will automatically select the best frequency on which to listen or transmit. Other useful services of modern computerized radio include automatic frequency-hopping and automatic dialing into a public telephone network.

Frequency-hopping is jargon for randomly changing a radio transmission frequency hundreds of times per second. Because the transmission frequency is never constant it is difficult for an eavesdropper to listen to secret conversations and so the technique is often used with military and police forces.

So many people are now transmitting and receiving on radio sets that interference is a constant problem. This is particularly true in large towns and cities where professional services must compete with each other for air space. To combat this, modern car radiotelephones have what is called automatic selective calling units. The person in the car presses a button or flicks a switch and the unit will listen to all the frequencies available and, when one becomes free, automatically call the appropriate telephone switching center. At the switching center an automatic receiver converts the radio signals into the signals used by ordinary telephones. When ready, the radio bleeps to inform the driver that his call has been connected. While this relatively expensive technique is currently of interest only to the military and police forces, it is beginning to spread to civilian markets and may be popular in the near future.

As all of these techniques become more common and microchip technology becomes less expensive for the manufacturer and the consumer, new applications like the wristwatch radio will cease to be merely an idea: by 1990 it could very well be a toy that people give to their children.

Antennas

Most communications signals are transmitted along cables or through the atmosphere. When sent through the atmosphere they are transmitted and received by devices called antennas.

In principle, all antennas function in the same way. For example, a transmitter causes alternating electric currents — currents that change direction rapidly back and forth — to flow in a piece of wire. Electromagnetic radio waves radiate outward from the wire like the ripples caused by dropping a stone into a pond. At some distance from the transmitter another piece of wire is connected to a listening device called a receiver that is able to detect minute amounts of current being generated in the wire by the transmitted wave. In the receiver special electronic circuits boost the weak signals to a usable strength. In this example, the pieces of wire are acting as antennas.

There are different antennas for different uses. One important factor in determining the shape and size of an antenna is the frequency of the radio signal. The higher the signal's frequency, the more important it is that nothing stands between the transmitting and receiving antennas. High-frequency radio signals, for example, can bounce repeatedly between the earth and a layer of the earth's atmosphere that begins fifty miles above its surface called the ionosphere. Passing with ease over tall buildings and mountains, even low-powered, high-frequency signals can be heard halfway around the world. High-frequency antennas range from huge tent-like structures to small metal loops placed on the tops of roofs.

At very-high- and ultrahigh-frequencies (VHF and UHF), radio waves travel in straight lines, covering a distance of about thirty to sixty miles. Although such waves will pass through some buildings and bounce around obstacles, the best reception is obtained when the receiving antenna is directly in line, or in line-of-sight, of the transmitting antenna. VHF and UHF frequencies are often used by mobile radio operators. To obtain maximum coverage between the base station and radio cars, the base station antennas are mounted on the tallest available building or on a hill. Such antennas and receivers are omnidirec-

A dish antenna points heavenward, tracking a communications satellite in orbit thousands of miles above the earth. The antenna transmits and receives microwave beams, rotating on its axis to remain focused accurately on the distant satellite as it drifts slightly in its orbit.

tional, which means they transmit and receive in all directions. A common type is a simple upright rod, sometimes with crosspieces. Car and some ship antennas are known as whip aerials because they are thin and flexible.

If a reflector is added, the antenna's transmitting and receiving performance becomes directional. This means that the antenna functions well in one direction and only partially in all others. One example is the yagi antenna — the flat, fish-bone structures seen on roof tops and used for color television reception in many homes.

Microwave antennas are usually dish-shaped. The microwave signal is passed through a transmitter in front, or on the side, of the dish and then bounces off the surface of the dish to form a concentrated beam of microwaves. If the beam is directed toward the sky at an oblique angle, parts of the beam will be turned back toward earth by the atmosphere and scattered forward far beyond the horizon, to be caught by antennas on the ground. Because this technique utilizes the lower layer of the earth's atmosphere called the troposphere, it is known as tropospheric scatter, or troposcatter for short. Some troposcatter antennas resemble huge curved billboards and so are known as billboard antennas.

Satellite communications use signals at microwave frequencies radiated from dish antennas that point toward the satellite. In their early days these dishes were about ninety feet in diameter and in heavy rainfall or a downpour would quickly collect rain that interfered with transmissions. As electronics advance, satellite systems become more powerful, higher frequencies can be used and dish antennas can be smaller. One result will be widespread use of satellite dishes for direct reception into homes.

Below are some common antennas. The yagi operates in one direction only. Omnidirectional and whip are used in mobile radio systems. A ferrite rod is used in radios and televisions. A rotatable periodic is used to cover a wide range of frequencies.

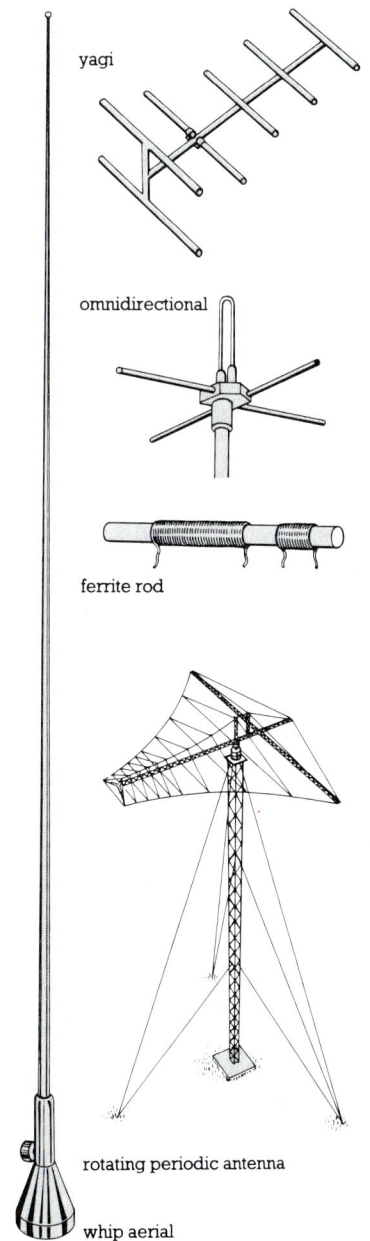

yagi

omnidirectional

ferrite rod

rotating periodic antenna

whip aerial

Amateur and Mobile Radio

For many people radio holds such a fascination that they have their own equipment to transmit and receive radio messages. After passing a simple examination, operators can apply for a license to transmit. A call sign made up of letters and numbers is allocated to each person who receives a license, and is used to identify each transmission and may indicate its origin as well. Any authorized frequency can be used in any of the international radio wavebands, so an amateur can make radio contact with any one of thousands of other enthusiasts.

As radio components become increasingly sophisticated and high-powered, many amateurs are able to transmit over greater and greater distances. Most radio enthusiasts use radio signals known as shortwaves. Shortwaves have wavelengths approximately thirty to 300 feet and are in the high-frequency waveband. Shortwave radios emit signals that cannot escape the earth's atmosphere. Instead, they travel thousands of miles by bouncing back and forth between the earth's surface and the ionosphere.

Less ambitious radio transmissions can be made by people who have their own VHF sets. VHF waves travel in straight lines and so the earth's curvature limits their range to about sixty miles. There is an advantage to VHF's short range, however. As long as

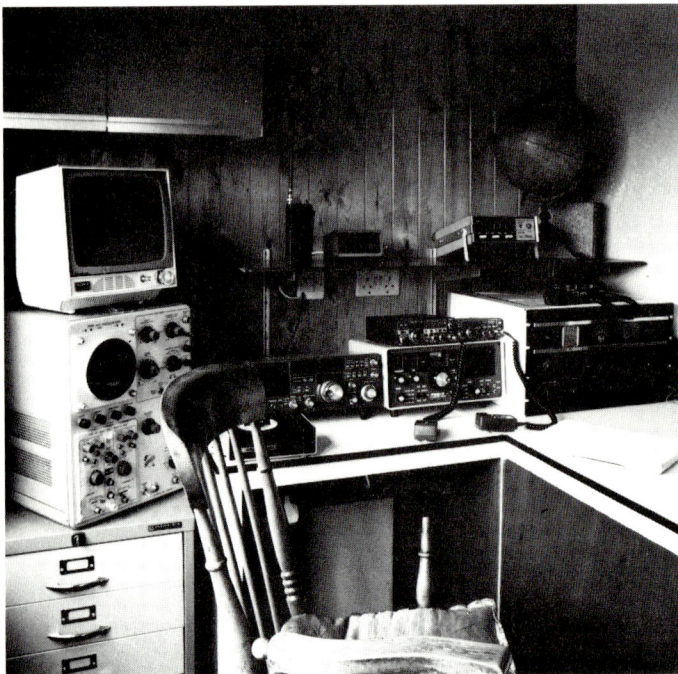

The equipment used by an amateur "ham" radio operator can range from the most basic to very complex. The type here shows an oscilloscope, left, and transmitters and receivers.

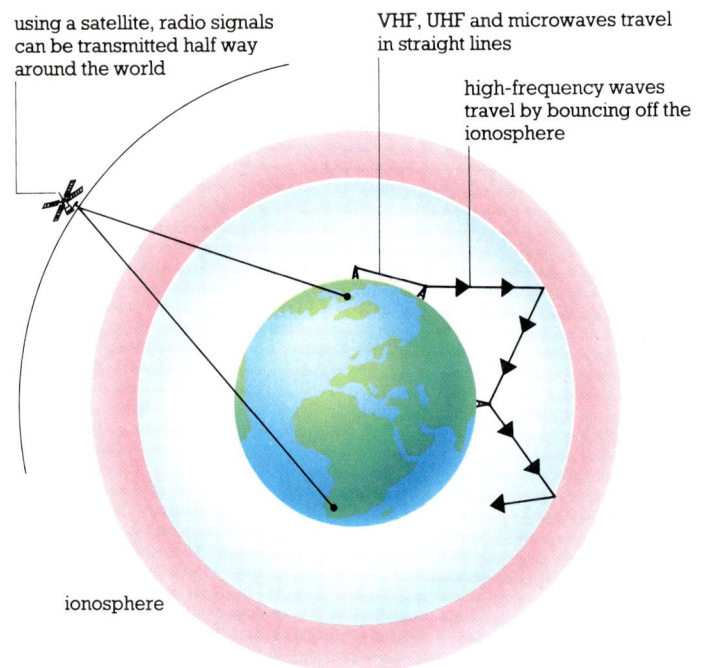

using a satellite, radio signals can be transmitted half way around the world

VHF, UHF and microwaves travel in straight lines

high-frequency waves travel by bouncing off the ionosphere

ionosphere

This diagram shows how radio waves can be transmitted to cover short or great distances. By using a satellite or by bouncing the signals off the ionosphere, great distances can be covered.

their transmitters are all far enough apart, many people can use the same waveband without disturbing one another. As a result, VHF is used by both professional broadcasters and mobile radio operators.

Today, with microchips that contain hundreds of thousands of transistors, and improved radio components and design, some enthusiasts use advanced radio sets operating on high frequencies of about three billion cycles per second — the microwave bands. These operators talk to others over great distances by bouncing messages off the moon or by using satellites reserved especially for their use. Such enthusiasts are now working toward developing radio teleprinters and the exchange of computer information by using special radio codes.

A form of amateur and mobile radio popular in the United States and becoming legalized in other countries is citizens band radio, commonly known as CB. CB was originally used by long-distance truck drivers in the United States while on long cross-country routes and has a unique slang vocabulary that many users have adopted. CB is not only a source of entertainment for long-distance truckers — it is also a useful means of communicating important information such as road and weather conditions, danger spots, highway patrols, and emergencies.

CB suffers one drawback, however. Because the legal CB frequency band is so narrow, users' signals often interfere with each other. To avoid this, some users add a booster to give them more power and therefore a stronger signal. Boosting can, however, make the transmission more powerful than the legal limit, which means the user will get through but may spoil the conversations of other users. The countries now allowing CB usually have a legal limit on transmitting power. Even if the police are unable to catch a CB-user ignoring these regulations, other law-abiding users will, and they have their own ways of controlling unruly operators. They can jam the illegal transmissions by playing loud music on or near the offender's frequency, or they can verbally reprimand him over the air so everyone can hear.

Military radios such as the one shown here are vital in enabling soldiers in the field to keep in constant contact with their headquarters and to keep in touch with each other.

Microwave

When commercial radio broadcasting began in 1920 the medium-frequency (MF) radio signals then used could travel up to 1000 miles. Soon after, amateur shortwave radio operators discovered high-frequency (HF) radio bands. Because waves in this frequency can travel long distances, it was used for worldwide communication. Very-high-frequency (VHF) soon followed, but was unable to travel around the curvature of the earth. Because of this and because the shortwave band was already overcrowded, engineers decided to use VHF for television broadcasting. However, VHF soon filled to capacity and engineers turned to UHF for television broadcasting. Today VHF is used largely for mobile communication links between aircraft and ground stations, and UHF for television broadcasting. As existing wavebands became more crowded, even higher frequencies had to be found, leading to the use of the microwave band.

Microwaves are so short that many billions pass a given point in less than a second. One benefit of using microwaves for communications is that they can be focused into a beam as narrow as the light beam of a flashlight. Transmitted between the dish antennas

High over San Francisco, an engineer inspects a microwave antenna.

of ground stations, microwaves can thus carry messages extremely efficiently: a tower 100 feet high, for example, can beam information over 100 miles to a dish antenna several feet in diameter. In addition, they require little transmitting power and can be used to send large quantities of data.

Television pictures contain tens of millions of bits of information per second, while a telephone conversation requires many fewer bits of information. Advanced commercial microwave transmitters can carry about 100 billion bits of information per second, which adds up to a dozen television channels and about 10,000 telephone conversations simultaneously.

The technology of transistors has spawned the use of small-scale microwave generators. Police and ambulances use microwave radiotelephones to link their vital control centers. Big and small companies find them to be an economical means of sending data between factories. When computer centers are linked by microwave it is possible to transmit large amounts of data among a number of machines in dif-

This microwave antenna in Oman, on the southeast coast of the Arabian peninsula, is connected to a solar-powered transmitter.

This diagram shows how a microwave landing system (MLS) guides an airplane on its approach to an airport. The MLS antenna transmits a scanning beam that moves back and forth and up and down across the runway. A plane entering a given area will block the microwaves, sending signals to a computer that can determine the position of the aircraft.

outgoing aircraft

incoming aircraft

MLS antenna

scanning beams

ferent locations, all at the same time.

Navigators often use a microwave communication system called radar to plot their own or another vessel's position. This is done by sweeping a narrow beam of microwave energy across a volume of space. The beam sweeps regularly, scanning at predicted times like the sweeping light of a lighthouse. A microwave beam can be so finely focused that it can indicate position to within inches, up to twenty miles away from the transitter.

Crisscrossing microwave beams are used to guide space shuttles in their last few minutes before touchdown, and similar systems are being developed for use at airports. Microwave guidance is more precise than radio and can make flying safer.

The concentrated beam of a microwave can be easily obstructed. While this is a disadvantage to the communicator it has been put to good use in burglar alarm systems. Any object that moves into the path of the microwave beam instantly triggers an alarm. Transistorized burglar systems are small enough to be mounted on fences or hidden alongside windows and doors. These are examples of how microwave-furnished information is contributing to everyday life.

Secure Communications

Most people do not like other people to eavesdrop on their conversations, but for some people secrecy in their communications is more than a matter of privacy — it is vitally important. Heads of governments and military leaders obviously need to ensure that their signals remain secret for reasons of national security. Military communicators devote more effort to the science of cryptography — writing and decoding secret messages — than anyone else.

Secret printed messages have been sent in code for years. A crossword anagram is an example of a simple type of printed code. If a person understands the coding method used, he can decode the message quite easily. If he uses a small, powerful computer, he can do it even more quickly and will be able to decipher even the most complexly coded message.

Securing vocal communications is another process. In civilian life, one method for making a telephone conversation secure is to use what is known as a scrambler. An electronic device in the mouthpiece of a telephone jumbles up the voice of the person talking and a similar device at the receiver's end unscrambles it. Anyone determined to intercept a message, however, can record the conversation and decode it at his leisure.

On the battlefield where important messages are sent by radio, the enemy can be prevented from communicating by jamming his signal, that is, by

The Enigma machine (below) was used in World War II to secure messages. In the picture at right, a soldier encodes a message on an Enigma keyboard for transmission by the radio operator to his right. The transmitter (center left) was used to send coded signals to other field units.

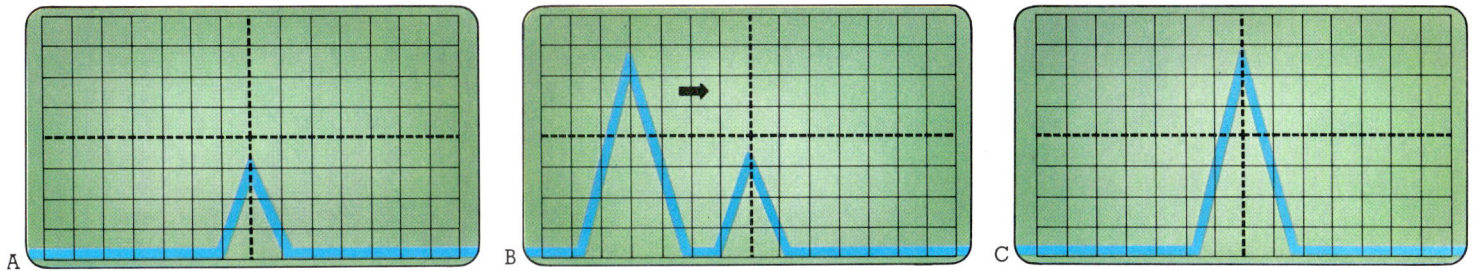

A   B   C

transmitting noise or music on or near the same frequency. The weakness of jamming is that everybody knows you are doing it and so they can easily locate and destroy your transmitter. Also, a great deal of power is required to interfere with, say, a radio station transmission. Often it is preferable to allow the enemy to transmit and then try to intercept his messages. One well-proven method of foiling such an attempt is simply to compress the signal being transmitted, a technique similar to recording a message at one speed and playing it back many times faster. Unless the eavesdropper is quick or keeps a continuous watch, he will miss the transmission. In practice, a message of several minutes can be compressed into a burst of sound lasting a few seconds.

The most modern technique is to change the frequency of the communication. With frequency-hopping a communication can be made almost completely secure. This technique involves transmitting a message over a spread of frequencies randomly selected by a microprocessor. Even the person sending the message does not know which frequencies his transmitter is using. The authorized receiver knows because the transmitter sends a special signal telling the receiver what the new frequency is. Friendly receivers tune into the chosen frequencies between bursts of information. As each burst lasts only a few thousandths of a second, the receivers must retune many hundreds of times per second. To successfully jam a frequency-hopping transmitter, it would be necessary to transmit noise continuously over all the wavelengths in use, which requires a great deal of power. This technique was designed for such organizations as the military services, which can afford expensive equipment, but is also used in civil communications. In the latter, frequency-hopping techniques can be used to avoid interference problems and to improve the quality of the signal that reaches the receiver.

This diagram is a visual representation of how a radio signal can be jammed. In A, a low-level, fixed frequency transmission is shown before interference. In B, a jamming signal — the higher peak — is shown moving in on the fixed transmission. In C, the fixed transmission has been "captured" by the jammer, creating unintelligible static on the receiver's end.

At the United States Strategic Air Command headquarters (below) in Omaha, Nebraska, some of the telephone calls received are first scrambled to ensure that the information they carry remains secure.

Radar

The countries that used radar in World War II had an advantage because they could detect approaching enemy ships and planes long before they were actually within sight. At that time, the enemy was revealed by radar as a series of coarse blips on a screen which gave the radar operator an approximate idea of the enemy's direction and distance. Today, radar is a far more precise tool.

The term radar comes from the phrase *RA*dio *D*irection *A*nd *R*anging. It is still used by armed forces to detect enemy ships and airplanes, but is also widely used for civilian purposes such as helping airplanes and ships to navigate.

Whatever their use, all radar systems rely on the reflection, or "echo", of electromagnetic waves. In a typical radar set, a transmitter and antenna send out short bursts of microwave energy focused into a narrow beam. A solid object in the path of the beam will reflect some of the energy back to the set where a receiver transforms these "echoes" into electronic signals. The signals then represent the object as a spot of light, called a "blip", on a visual display screen. With a rotating radar antenna, the beam can locate objects in any direction from the set.

The direction of a distant object can be determined by the direction of the microwave echoes; the distance is determined by the time between the transmission of the beam and the return of the echo. For example, if you were standing in a canyon and shouted towards one wall and then measured the time it took for the echo to return, this time, multiplied by the speed of sound and divided by two, would tell you the distance to the wall. (You must divide by two because the sound covers the distance twice, once as your shout and once as the echo.)

There are two main types of radar: the more common kind, called pulse radar, measures the distance and direction of an object by emitting timed bursts, or pulses, and then waiting for the echoes to return. The other kind of radar, called continuous wave, measures the speed of an object by transmitting an uninterrupted signal. Stationary objects return echoes at the

Huge spherical domes house rotating antennas at a NATO radar station in Yorkshire; England. This powerful early warning radar can detect approaching aircraft thousands of miles away.

frequency of the original beam, but echoes from moving objects have different frequencies. This principle can be used to measure the speed of an object in relation to the radar transmitter.

Airplane radar is of the pulse type, emitting several hundreds of bursts per second, each a millionth of a second long. The time between pulses is long enough for echoes to return from objects up to 300 miles away. The radar beam is transmitted ahead and to either side of the aircraft, gathering information in an arc ahead of its path.

Airport radar is also of the pulse type but uses larger antennas and generates more powerful pulses than those used by aircraft. These factors improve its sensitivity and allow it to scan in all directions.

The speed-trap radar used by police is continuous wave. The transmitted energy is modulated by varying the frequency of the radar beam either side of a predetermined level. When the beam bounces off a moving vehicle, it comes back at a slightly different frequency, or pitch. The change in pitch is caused by the movement of the vehicle. This effect was first noted by the Austrian physicist Christian Johann Doppler. The Doppler Effect, as it is now known, can be heard in any large city. Horns and police sirens, for example, seem to change their tone as they approach or move away from a stationary person.

Military radar for early warning of enemy attacks demands very skilled radar operators who must be able to recognize the difference between echoes from the earth's surface and those from aircraft and ships. To do this both pulse and Doppler methods are used. The operator can then calculate the direction and range of a target from the returned pulses, and the speed of the target from the Doppler shift in the frequency of the continuous wave.

In the radar room of a ship, operators examine a visual display unit for data identifying nearby vessels. A diagram below shows the steps involved in detecting a distant fighter plane by radar.

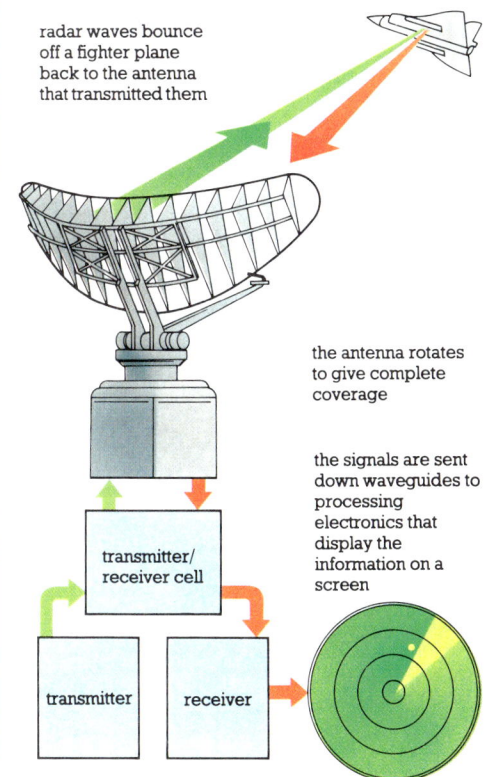

radar waves bounce off a fighter plane back to the antenna that transmitted them

the antenna rotates to give complete coverage

the signals are sent down waveguides to processing electronics that display the information on a screen

transmitter/ receiver cell

transmitter

receiver

Sonar

In the picture at right a helicopter is searching for submarines by dipping its sonar device sixty feet into the sea. Using sonar in this way makes it possible to hunt for submarines without exposing surface vessels to attack.

How is a signal sent through water? Most of the communications techniques used to transmit messages in the atmosphere are inefficient in water because radio waves are quickly absorbed. But pressure waves, or sound, travel easily in water over long distances. Ships and submarines, for example, can be heard miles away by lowering a microphone into the sea.

The word sonar is derived from *SO*und *NA*vigation and *R*anging. Sonar is widely used by submarines to detect ships — and in all ships to measure water depth. Sonar can be used for communication, but normally only between ships near one another.

Measuring water depth by sonar is straightforward. The sonar operates like a pulse radar, sending out sound waves from a ship's keel. The waves bounce off the seabed, returning to the ship as echoes. The time it takes for the echoes to return is translated into the distance between the ship and the seabed.

Sound waves travel less predictably in water than do radio waves in the atmosphere. Temperature variations and sea currents deflect them, weakening or

Blips on a sonar screen (below) show obstacles that the sonar sound waves have encountered.

Trailing a transmitter, a sonar-equipped ship (below) sends sound waves to the seabed that bounce back to a receiver

beneath the vessel. An airplane (left) gathers information by radio from a sonobuoy. The submarine (below) uses fixed radar.

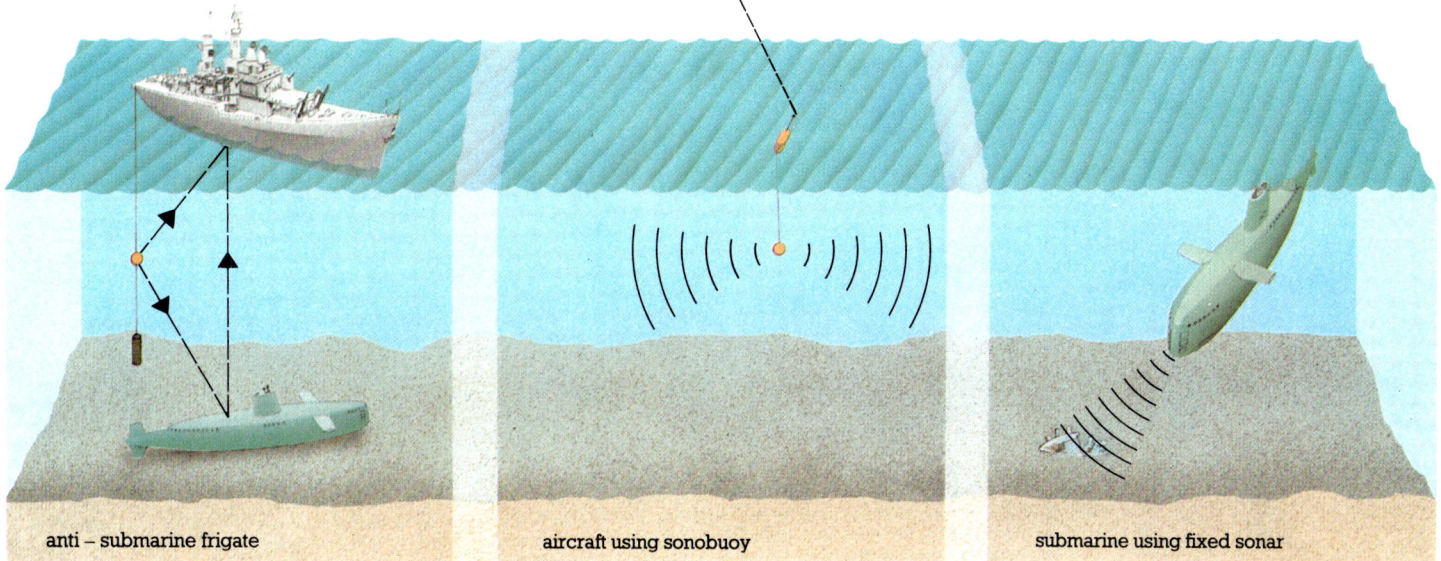

anti – submarine frigate   aircraft using sonobuoy   submarine using fixed sonar

amplifying the original noise level. This is not a problem for small depth-measuring sonar, but creates difficulties for engineers designing equipment used to determine the position of other vessels. Sea creatures such as whales complicate the problem because many communicate by making sounds, sometimes loud enough to mask the noise of even large ships.

There are two types of sonar used by vessels at sea. Passive sonar listens for sounds, while its counterpart, active sonar, emits pulses of sound which are reflected by objects below the surface of the sea. Passive sonar cannot reveal a ship's position nor indicate the distance of an object, only an approximate direction. Active sonar is used when distance or depth must be determined accurately.

Passive sonar is the sensor used today by many military submarines to detect surface targets. It can also be used by a warship that wants to avoid being sunk by a submarine. The ship can tow a network of passive sonars spread out at different positions and depths. Electrical circuits in the ship combine all the sounds detected by the sensors and build a picture of what is happening — including the presence of a submarine — below the surface of the water. The sensors must distinguish between noises created by the ship and sea sounds, especially if part of a convoy.

A military submarine trying to stalk a ship by using its own sonar detectors will not know if it has been detected. This is an advantage for the ship on the surface. If the net of sensors is wide enough and the sea not too rough, the submarine can be found with enough accuracy to destroy it.

Aircraft are often used to drop sonobuoys — small sonar sets with a collar to keep them afloat — into the ocean. A mechanism in the sonobuoy lowers a microphone hundreds of feet below the surface of the water and a transmitter sends the sounds picked up back to the aircraft. If an airplane drops a spread of buoys, it can survey many hundreds of miles.

Sonar sets can also be placed on the seabed and left for several years. Ships passing overhead can activate these listening posts to monitor the movements of all ships within that area of the ocean.

Sonar is often used in oil drilling. A receiving system is spread at regular intervals over an area of land or sea and an explosive is triggered either above or below ground. The sound waves that result from the explosion penetrate into the earth and come back at different amplitudes depending on the type of material they hit. These are then used to calculate the type of rock or mineral beneath the earth's surface, and whether oil is present.

Civil Satellites

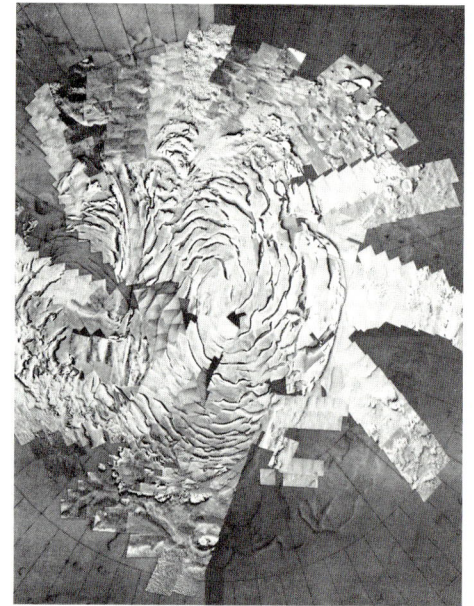

Satellites are able to take series of small pictures which, when put together can create a whole image. This "mosaic" photograph is of the North Pole of Mars, taken from the Viking II orbiter.

In a testing chamber (left) a satellite is tried out during simulations of the varied conditions it will have to endure: the shock of the launch, the searing heat of the sun during half of its orbit, and the deep cold of the far side of the earth.

Engineers and scientists on earth depend on microwaves to link themselves with satellites in orbit, using large, dish-shaped antennas to beam the waves upward from transmitters on the ground. To do this, the dish of the antenna focuses the microwaves into a narrow beam, concentrating the energy to be received by the satellite. Communications satellites use microwaves because they are able to penetrate the earth's atmosphere without being altered or distorted by normal atmospheric conditions.

To maintain the long-range communications link, microwave antennas on the ground must be able to follow the satellites as they orbit the earth. In low orbits of a few hundred miles above the earth's surface, satellites circle the globe once every one and one-half hours or so. Any one ground station can only communicate with such satellites for about twenty minutes at a time. However, when a satellite is placed 22,300 miles above earth, it completes an orbit in twenty-four hours, exactly the time it takes the earth to rotate on its axis. As a result, the satellite stays over one spot on the earth. All modern communications satellites are placed in this type of orbit, called geosynchronous orbit. A satellite in a geosynchronous orbit is in radio sight for about one-third of the earth's surface. Consequently, signals transmitted to the satellite by any one earth station can be transmitted back to earth to any other station up to 8000 miles away from the original transmitter.

A satellite in orbit will wander slightly due to the gravitational pull of the moon and planets, and so its position must be constantly corrected by engineers on earth. They do this by causing small rockets on the satellite to push it back into its correct orbit.

A satellite has to generate power continuously to operate its receivers and transmitters. It does this, in part, by collecting solar energy which is converted directly into electric power by solar cells. Even though the energy from the sun is massive, solar cells are inefficient, and the energy they produce must be used economically. For this reason the transmitter in the satellite is smaller than one might expect for a link of 22,300 miles. The signals are directed from a small dish antenna on the satellite to a large one on the earth where they are amplified to a practicable level for use by ground personnel.

A person making a telephone call transmitted by satellite often notices a slight delay in the conversa-

This diagram shows the position of three MARISAT (Maritime Satellites) satellites in their geosynchronous orbits over the earth. MARISAT was created to give international aid to ships positioned all over the world. It transmits and receives a continuous flow of information to keep ships in touch with the shore and with other ships.

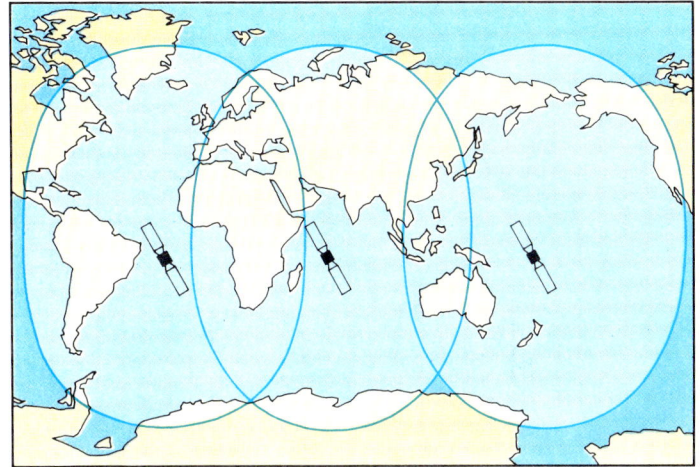

tion. This is because the microwave signal takes about a quarter of a second to travel between the earth's surface to the receiver in geosynchronous orbit and back to earth again. Although frustrating to the person making the call, considering how far the signal is traveling, this time-lag is not of consequence.

Most satellites in orbit are dedicated to serve particular areas, for example, North America or Europe. There are also international systems made of chains of satellites that can provide communication links between up to 109 countries.

One goal of current research in satellite communications is to provide sufficient energy from an orbiting satellite to allow people on the ground to use small, inexpensive receivers with antennas only two or three feet in diameter. When this goal is reached— probably by 1985— television transmissions will be beamed directly into homes over wide areas by using dish antennas. This method will be much cheaper than continuing to use the many large ground stations needed at present. It will also be more efficient than ordinary television transmission from the ground because there will be no masking effect created by mountains, forests, or towns.

Military Satellites

Military communications satellites allow generals in their headquarters to talk to their troops at any time, wherever they may be. Satellites are well suited to military use because they can transmit microwave energy focused into a narrow beam that is not easily detected by the enemy. Satellite communication is also less prone to interference and disturbance than telephone and radio communications transmitted within the earth's atmosphere.

Sailors have used satellite navigation systems increasingly since the early 1960s. Just as stars have provided accurate markers for navigators for centuries, so satellites are a type of star that can tell naval commanders where they are. For ships at sea, special military navigational satellites can be used several times per day for an hour or so at a time, at any location, to determine position with a high degree of accuracy.

Perhaps the ultimate expression of this is a network of eighteen satellites currently being developed. Called the Navstar Global Positioning System, it has been developed by the United States primarily for its military forces, but it is available for civilian use as well. With a highly portable ground receiver — a box slightly larger than a hand calculator — a soldier or sailor will be able to find his position anywhere on the earth's surface to within

This is a satellite picture of the North Pole. Such pictures are often used by military personnel.

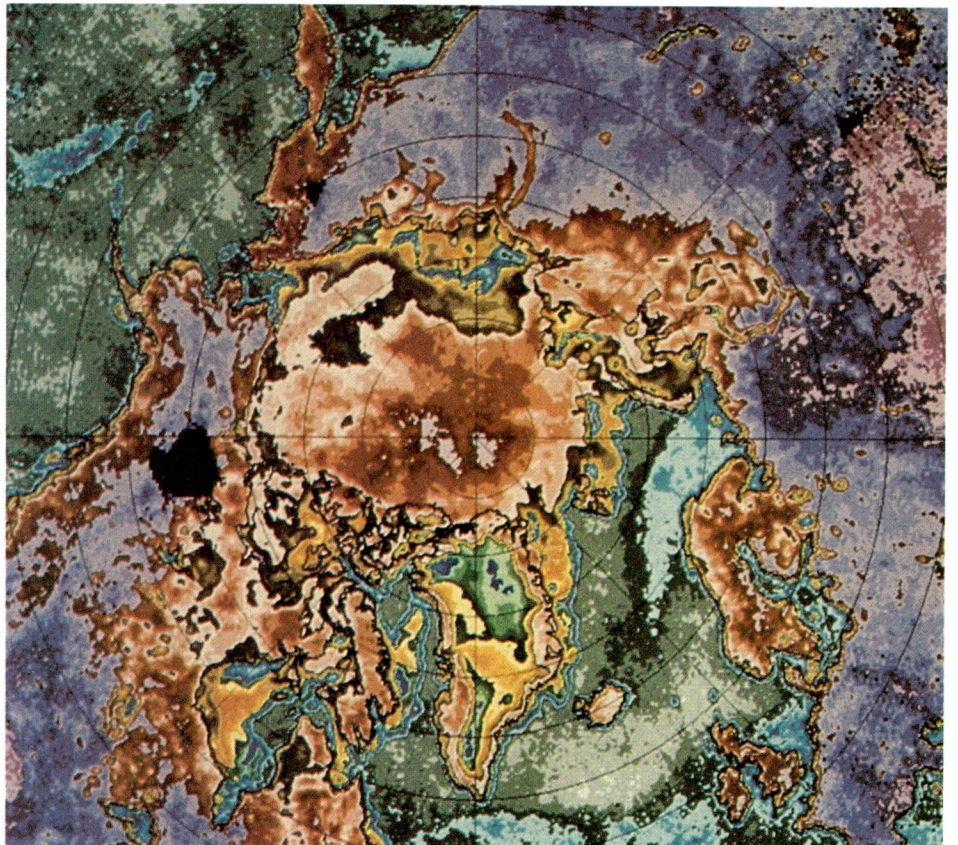

fifty-two feet. Another highly secret satellite system has an even more remarkable capacity. Using a set the size of a soldier's backpack, a user will be able to fix the position of an enemy ship or tank to within sixteen feet — anywhere on the earth's surface, at any time of day.

Military satellites can alert commanders to military activity taking place anywhere on the earth's surface. There are special satellites that can, for example, detect the infrared glare created when a missile is launched, spot nuclear explosions, or track vessels at sea. Other satellites can be used to eavesdrop on other countries' communications links.

Reconnaissance satellites take photographs of security areas inaccessible by land or sea. These satellites send information back to earth in a microwave beam that, when processed, provides a picture made of tiny squares like a mosaic. The size of the mosaic is normally compact enough to show fairly large-scale activity. Pictures taken at different times can be automatically compared using computerized pattern-recognition machines. A difference in the patterns on a coastline, for example, might betray the presence of a secret new naval base. Areas of the mosaic that are of particular interest can be electronically scanned to produce more detailed pictures called enhanced-image pictures. The pictures can also be colored artificially to highlight certain aspects.

If such surveillance is not adequate in itself, there are also satellites that take photographs using conventional cameras and film. These have a special telephoto lens capable of spotting objects as small as one foot across from over 100 miles above the earth's surface. The satellite is brought down with the photographs after a week, or it may eject a small package that carries the exposed film to earth at daily intervals. The capsule can be parachuted onto a desert or retrieved as it falls by specially equipped aircraft. This method is much more expensive and time-consuming than the mosaic technique and is used only when very high quality photographs are required for detailed information.

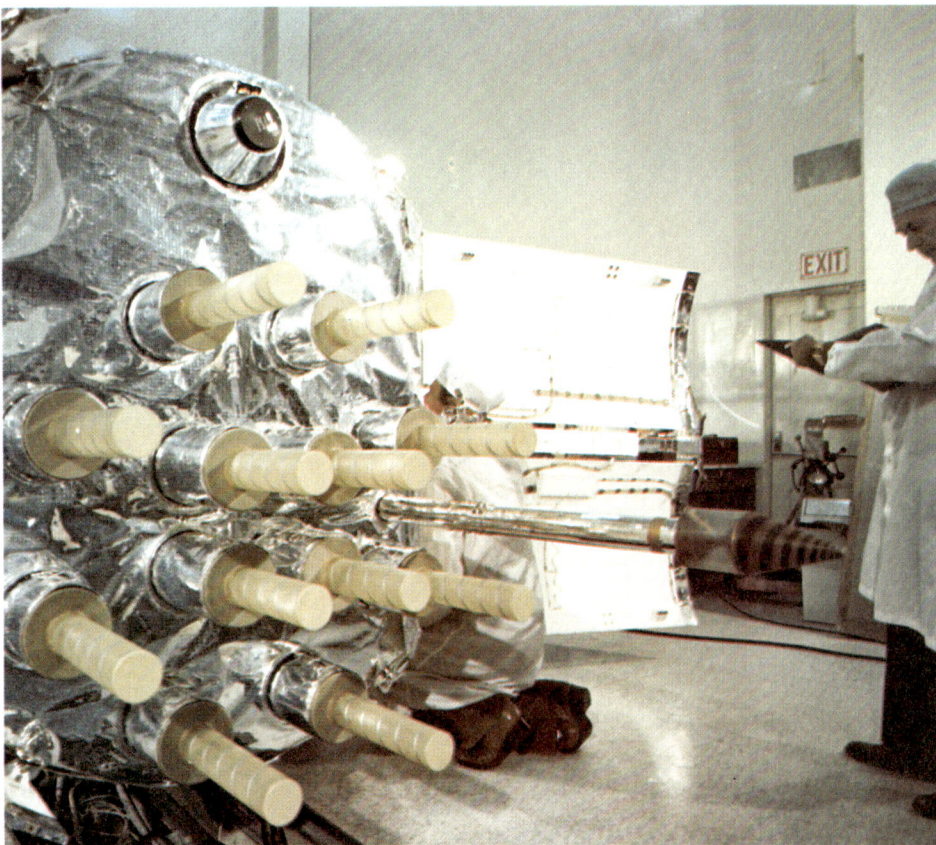

The diagram above shows the path of the eighteen NAVSTAR satellites that will circle the globe, providing navigational aid. At left is a picture of engineers checking out a NAVSTAR satellite prior to packaging it for shipment to the United States Air Force launch site in California.

Satellites as Tools

A satellite in orbit has a broader view of activities taking place on the earth's surface than any surveillance device previously devised by man. With the advent of the reusable space shuttle, which can inexpensively launch and retrieve satellites in space, satellites will be practical for more and more information-gathering services.

Most satellites are not manned and must be told what to do by people on earth. They must also be able to send the information they gather from space back to earth. For these reasons, sophisticated telecommunications techniques are needed.

Before the advent of satellite technology, man's knowledge of the earth was limited to aerial photography and theorizing. One important job satellites are now used for is the making of photographic maps of the earth's surface. The photographs taken so far have vastly increased man's knowledge of the geography and topography (details of geographic features) of remote areas of the world.

A new breed of sensors is extending the role of satellites beyond simple geographic and topographic mapping. These sensors can help man explore remote geological formations on the earth's surface that might contain minerals or oil. Such satellites can also give an overview of agricultural areas in order to check the extent of crop damage caused by disease or to monitor the effects of the weather on the growth of new plant species.

Weather satellites are now commonplace, and photographs taken from satellites of cloud formations and air movements on earth are often shown on television with the weather forecast. Instruments carried aboard weather satellites can measure changes in temperature and humidity on the earth. They can scan large areas of the earth's surface and provide weather information about remote areas where there are no weather stations. With the data provided by

A complete weather system over North and Central America appears in a picture taken by a weather satellite. The white lines and numbers are measures of the relative humidity in the atmosphere.

This is Meteosat, Europe's main weather satellite. It provides valuable information for meteorologists.

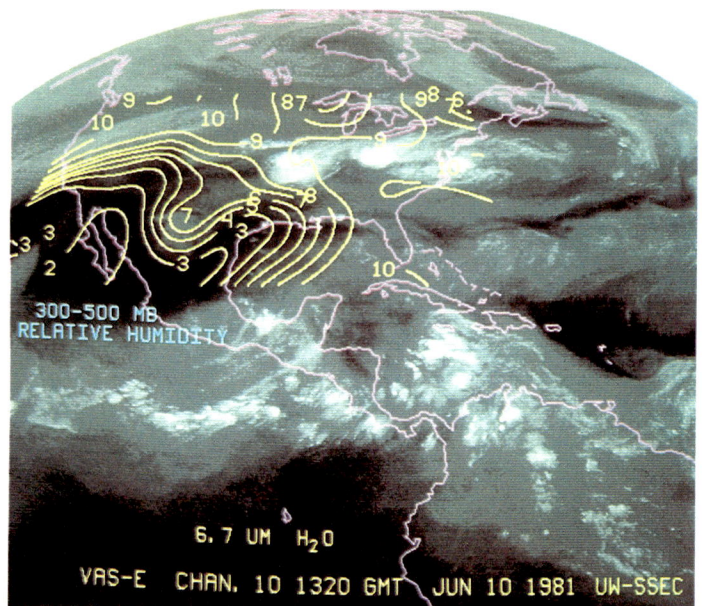

300-500 MB RELATIVE HUMIDITY

6.7 UM H2O

VAS-E CHAN. 10 1320 GMT JUN 10 1981 UW-SSEC

A photograph of the Libyan desert (left) taken by satellite is made up of a number of small views and artificially colored to enhance topographical details.

This is a satellite photograph of the Rhub Al Khali, Saudi Arabia. Computers are used to add color to show rock formations.

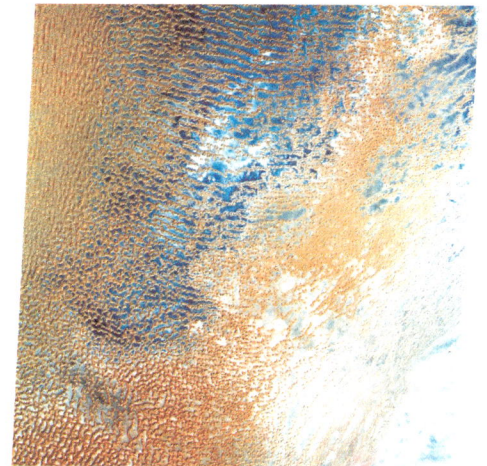

satellites, meteorologists can now prepare a master cloud map of the world every twenty-four hours.

Satellites carrying X-ray and ultraviolet telescopes allow scientists to detect radiation emanating from nearby planets and the remote parts of the galaxy. Without satellites it was very difficult to study these signals because the earth's atmosphere blocks them from ground observatories. By the mid-1980s, scientists hope also to put an optical telescope in orbit, which will be able to take pictures of objects beyond the range of the largest optical telescopes on earth.

Without microwave links between orbiting satellites and the earth, none of this information would be available. Communication links are vital, especially for craft that travel beyond the earth's orbit. Some of these are on their way to the outermost planets of our solar system and are expected to continue out through our galaxy, the Milky Way. The communications systems of such vehicles transmit beamed microwaves similar to those of orbiting satellites, but with larger antennas. Large antennas are needed because of the great distance over which information must be transmitted and received accurately.

Interplanetary spacecraft that sent back pictures of Jupiter and Saturn, for example, were so far away that although microwave signals travel at the speed of light — 186,000 miles per second — it took as long as one and one-half hours for the pictures to reach earth. At such distances from the sun, satellites cannot use solar energy to recharge their batteries and must carry a small nuclear power plant that is able to supply electric energy.

From the interplanetary travels of NASA's Voyagers and others over the last few years, scientists have learned many new things about our solar system, such as the composition of the planets and their similarities to each other and to earth. Engineers are now working on communications techniques which will enable us to send information from even deeper in space. If there is any form of life elsewhere in our galaxy, it may be these hoped-for communications breakthroughs that enable us to discover it.

45

Communicating with Light

Like electricity, light is a form of energy that travels through space as electromagnetic waves. In 1905 the German-born physicist Albert Einstein established that light is actually a stream of separate bundles of energy, called photons, that have characteristics of both particles and waves. Photons travel in a vacuum at the speed of light. As far as scientists know, nothing in the universe travels faster than the speed of light.

Light occupies only a small segment of the entire electromagnetic radiation spectrum. Light usually refers to the visible part of this segment, but there is

Lighthouses, in use since ancient Rome, still play a vital role in the safety of ships at sea.

also infrared light at one end of the segment and ultraviolet light at the other, neither of which our eyes are able to perceive.

Ancient man did not know what we now know about the nature of light. He did, however, realize that it could be used to communicate effectively. A bonfire, for example, lit at the top of the hill could be used to warn of danger. In those days communication was simple; the fire communicated one clear signal or message and no other. Because man needed to send more complicated types of information, he soon invented ways to transmit a number of messages with one light source. For example, he could wave a flaming torch from side to side to say one thing, or up and down to say something else. By combining different signals in this way, early man learned to transmit more information — but only if the receiver was close enough to see the burning torch.

For centuries, fire proved an invaluable means of communication. In 1588 Sir Walter Raleigh learned of the approach of the Spanish Armada when scouts lit a series of beacons. Although fires were effective at night, however, they were difficult to spot in daylight and were limited in the distance over which they could send messages.

Perhaps it was the glint of the sun on a spear or a sword that sparked off the idea that reflected sunlight could be used to communicate. Wherever the idea came from, it led to the invention of the heliograph, a device employing a mirror or piece of polished metal to reflect the rays of the sun. The heliograph had an important advantage over fire: it could be flashed quickly, allowing people to use sophisticated codes to send many different messages.

The heliograph was more effective than fire, but it was useless at night and on cloudy days. The nineteenth century invention of flares afforded one solution, an improved way to communicate by fire. Flares burn brighter than torches and can be fired

light

electrical power
units

linked chain

counterweight

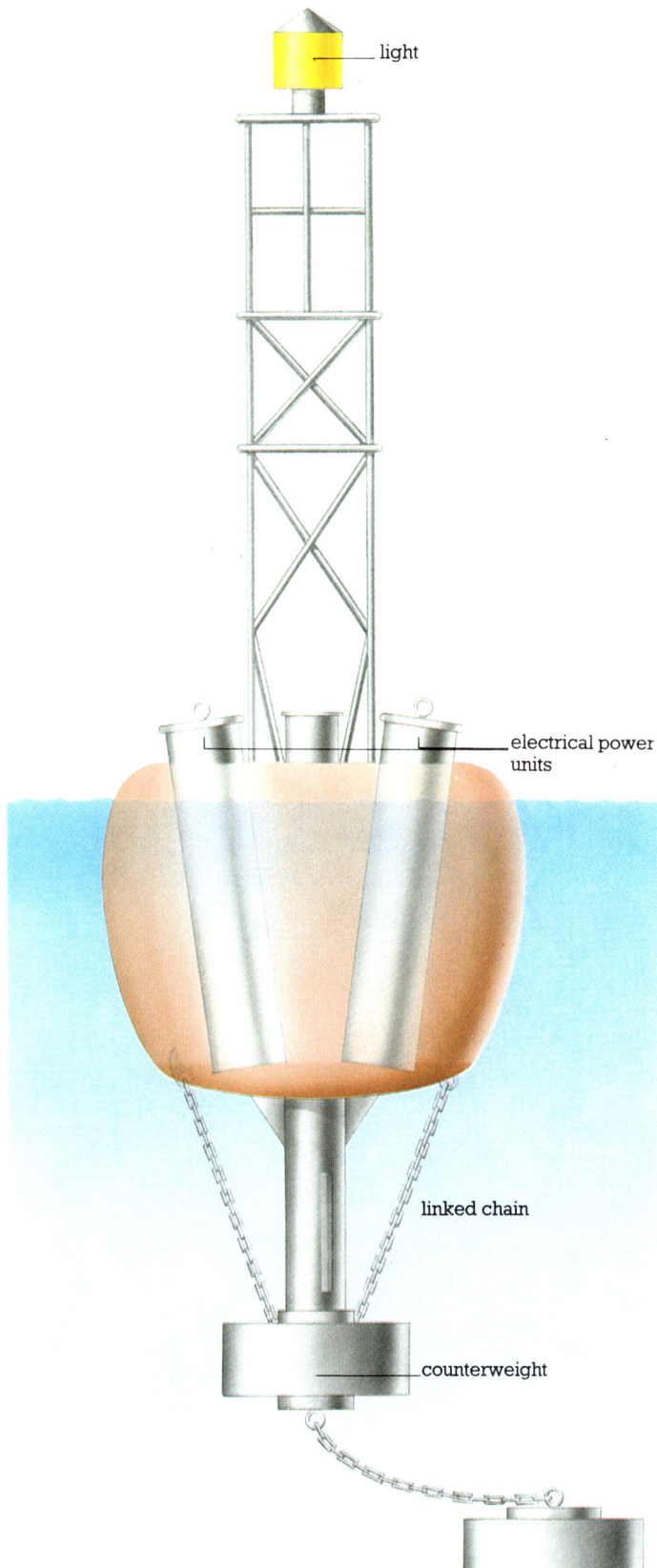

into the air to be seen at greater distances. Chemicals enable flares to burn in a variety of colors, so that yachtsmen, for example, can send messages in a code based on the colors. Flares are still used today.

With the invention of the electric light, man had for the first time a source of light that was powerful and easy to control. He could use electric light during both day and night to communicate over long distances. Even today military ships use powerful searchlights with shutters to vary a beam of light. Called Aldis lamps, these permit signalers to send messages in Morse code. Sometimes they communicate messages over the horizon by reflecting their light from the bottom of low clouds.

Although it has been greatly overshadowed in this century by electronics, light has continued to be an important medium of communication. Despite the prevalence of radar and other modern navigational aids, for example, sailors still rely on lighthouses to avoid the hazards of navigating at sea. Light now seems sure to enjoy a comeback in the form of fiber optics — transmitting laser light signals long distances through fine filaments of glass.

A buoy (left) carries its own power units to keep its beacon burning. More sophisticated buoys can flash lights of different colors and patterns. Buoys are often used to mark sea lanes.

When radio silence must be maintained, an Aldis lamp (below) can send messages in Morse code.

Fiber Optics

This engineer is making optical fibers from heated glass. Optical fibers are commonly made in lengths of about five miles.

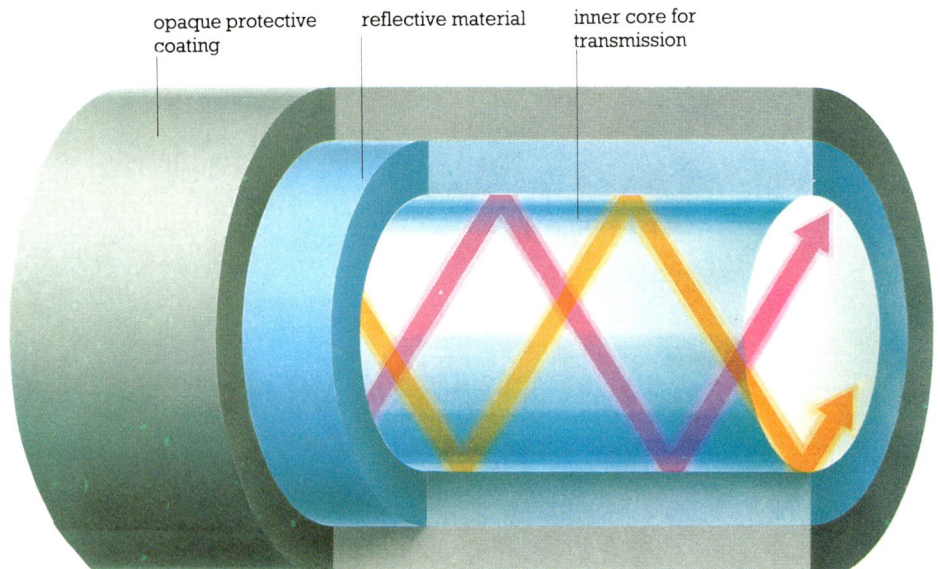

opaque protective coating

reflective material

inner core for transmission

In this diagram, light rays are shown bouncing down an optical fiber. Optical fibers transmit light in pulses. Each of these pulses

may contain several rays. Different fibers have different uses, for example, for long distance transmission.

For a long time scientists have been aware that a beam of light can transmit information more efficiently than electricity because of its high frequency. For example, the frequency of light is about 100,000 times greater than the highest radio frequency. It was not until 1970, however, that engineers finally devised a glass fiber that could carry a light signal far enough to be practical. Their discovery led to a revolution in communications. Not only may optical fiber cables ultimately replace the copper wires now widely used, but all communications may move away from using electrons as a means of communication, to using particles of light.

Optical fibers are made from glass 100 times more transparent than the glass in a window, arranged in an inner and outer layer called the core and the cladding. Light travels a little more slowly through the core than through the cladding. As a result, light rays entering the core at a shallow angle will bounce back

and forth off the inside surface of the cladding all along the length of the fiber, in accordance with a principle known as total internal reflection. The same principle applies when you look from underwater (through which light travels relatively slowly) up to the sky. If you look straight up, the surface of the water will seem transparent and you can see the sky. But if you look up at a shallow angle, the surface will be like a mirror: you will see nothing but light from underwater bounced back at you.

The source of light for an optical signal is either a tiny laser or a light-emitting diode, or LED, which is a type of transistor. Lasers are more powerful but cost more than LEDs and do not last as long. At the receiving end, the light enters a detector that converts the light into an electric signal.

Optical fibers have many advantages over conventional copper wires. A single optical fiber now in use can carry as many as 672 telephone conversations

Laser light, captured inside an optical fiber as thin as a human hair, can carry 672 telephone conversations.

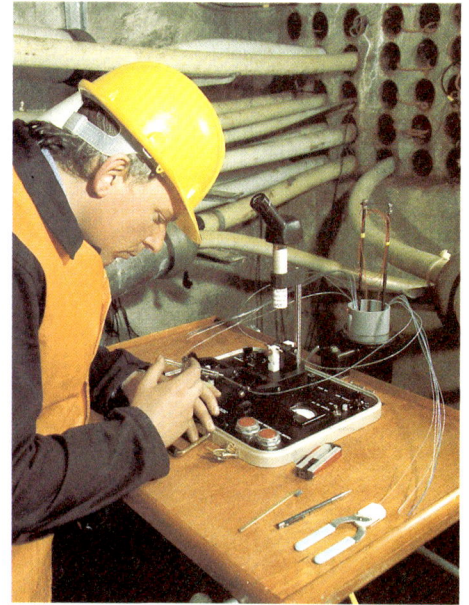

An engineer splices together optical fibers. This job requires special equipment; a bad joint means a poor signal.

compared with the twenty-four carried on a pair of copper wires. The capacity of fiber optic cables makes them the perfect medium for such services as telephone, cable television, stereo, cable radio, viewdata, and high-speed computer data.

A light signal sent down a glass fiber does not fade as quickly as an electric signal sent down a copper wire. While copper cables need repeaters every mile or so to boost fading signals, the optical fiber cables now being developed need repeaters only about every fourteen miles. Engineers have sent signals over 100 miles through fibers before repeating, but only in the laboratory.

Optical fiber cables are much lighter and thinner — about the thickness of a human hair — than copper ones, which makes them ideal for the crowded underground cable ducts of many large cities. They are also immune to the electrical interference that plagues copper cables — the buzzing, clicking, and humming audible on many phone calls.

Initially, optical fiber cables will be used mainly for what they are best at — connecting large switching systems where there is a heavy volume of telephone traffic. In addition to handling telephone conversations between people, optical fibers are ideal for transmitting quantities of data between computers.

Many countries, including the United States, Britain and Japan, have already installed optical fibers in local telephone networks, and some are also using fiber optic systems in their telecommunications networks. This is a step toward the day when every home will have a single fiber optic cable carrying telephone calls, television programs, facsimile signals, quadraphonic hi-fi — and even signals from automatic devices that read the gas and electricity meters. Such optical networks are already being installed in experimental projects and their widespread use may be just around the corner.

Integrated Optics

While fiber optics has proved that transmitting information by light has many advantages, at present optical transmissions still rely on electronic devices for essential functions like switching and amplifying. When a light signal carrying a telephone conversation has to be routed through a telephone switching office, it must first be converted into an electric signal, amplified and rerouted electronically, and, finally, converted back into light for the next leg of the transmission. In the near future, however, light wave systems will be able to perform these and other functions which now must be done electronically. The circuits that will make this possible are called integrated optical circuits (IOCs). They are called this because, as in integrated electronic circuits, all the components of the circuit will be part of a system formed on a single base material called a substrate. IOCs will direct a flow of light rather than electricity.

Like an optical fiber, the IOC relies on the principle of total internal reflection. A typical optical circuit of the future will consist of elements such as gallium or indium applied in pure crystal layers to build up a film forty millionths of an inch thick over a substrate of glass. Total internal reflection occurs because light moves more slowly through the film than through either the substrate on one side or the air on the

The engineer in this picture is testing integrated optical circuits using light rather than electricity. These circuits perform tasks such as switching and amplifying, and operate 10 to 100 times faster than the electronic circuits presently in use.

other. As a result, whenever the light ray strikes the boundary between the film and the substrate, or between the film and the air, it is reflected back into the film. The ray thus follows a zig-zag path, bouncing between the upper and lower inside surfaces of the film. A device that traps electromagnetic waves internally so that they will follow a desired path is called a waveguide. An optical fiber is one kind of waveguide; the kind used in an IOC is another, and is called a thin-film waveguide.

An IOC controls light waves by means of devices such as lasers, prisms, lenses and modulators (which impress information on the light wave). The simplest devices are the prisms and lenses, which refract (bend) or focus light just as ordinary lenses and prisms do. They can be made by creating a carefully shaped thickening in the waveguide. A triangular thickening makes a prism; a curved thickening makes a lens.

One possible type of modulator for an IOC is a device called the magneto-optical switch. This device takes advantage of the fact that light waves are electromagnetic waves, composed of oscillating electric and magnetic fields. Under certain circumstances, their electric and magnetic parts will be changed when the light wave passes through a magnetic field.

The magneto-optical switch is connected to an electric circuit that turns current on and off in accordance with, say, digital information sent from one computer to another. The on-and-off current turns a magnetic field in the switch on and off. If laser light passes through the switch when it is unmagnetized, the orientation of the light wave is unaffected. When the switch is magnetized, however, the orientation of the wave's electric and magnetic parts will be changed when the wave passes through the switch. If the light then passes through a polarized screen — a device that blocks light waves oriented in one direction and passes those oriented in the other — the once continuous light wave is converted into on-and-off pulses, corresponding to the computer's digital pulse. This pulse-modulated light wave now carries information and can be sent through a fiber optic cable.

Until recently, scientists have had less success inventing a suitable optical laser. Such tiny lasers were too short-lived or too easily affected by minor changes in temperatures to be practical. However, recently invented optical lasers can function at normal room temperatures and seem likely to run a million hours or more.

The major challenge still facing integrated optics is to reduce the power requirements of IOCs. At present the circuits need more power than they can withstand without melting. However, when the remaining obstacles are finally removed, integrated optics will be an important step for communications.

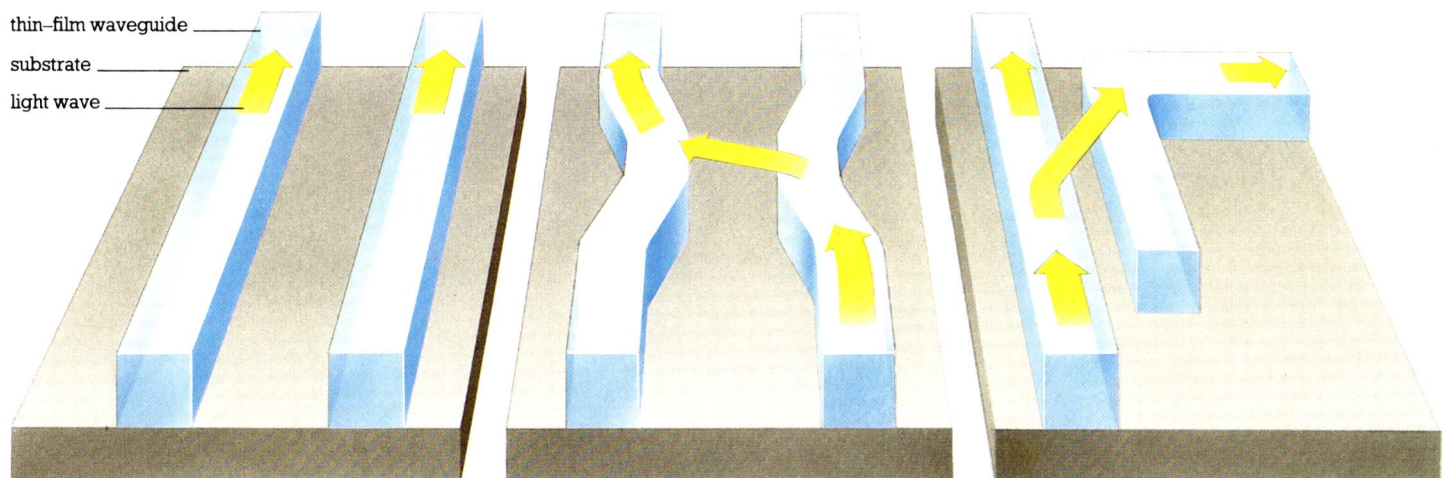

thin–film waveguide
substrate
light wave

These three highly magnified diagrams show how switching in optical microchips will be carried out by light rather than electricity. Left: light passes through two thin-film waveguides. Center: two waveguides are positioned about 2/10,000 of an inch apart, enabling light from one to switch into its neighbor. Right: voltage applied across both waveguides creates an electric field, causing only some of the light to pass over.

Traditional Electronics

In 1883, while experimenting with his recent invention the electric light bulb, Thomas Edison accidentally discovered what was to become the fundamental principle of electronics for more than fifty years. Sealing a metal plate inside one of his new electric light bulbs, along with the bulb's filament, Edison connected the plate to a battery. He discovered that a current began to flow from the hot, glowing filament to the plate. Moreover, the current flowed only when the plate was connected to the battery's positive terminal; when the plate was switched to the negative terminal, the current stopped.

Around the turn of the century the British scientist J. J. Thomson explained the mysterious "Edison effect" as the passage of electrons from the hot filament to the positively charged plate. The heat applied to the filament increased the energy of the electrons in the metal and caused some of them to break out from the filament's surface into the vacuum of the bulb where they were drawn to the positively charged plate. Thomson called the phenomenon thermionic emission, after *therme*, the Greek word for heat, and ion, the scientific term for atoms which have acquired a positive or negative charge by gaining or losing an electron.

In 1904, the British engineer John Ambrose Fleming applied this phenomenon to the field of communications by sealing two metal plates, called electrodes, into a glass tube from which almost all air had been evacuated. Using a battery, he heated one electrode — the cathode — and placed the tube in the circuit of a radiotelegraph receiver. As the incoming radio waves fluctuated from positive to negative, so did the charge on the unheated electrode — the anode — but because of the Edison effect, current would flow through the tube only when the anode was positive. As a result, the tube allowed only half of the wave to pass through the circuit — converting the positive and negative fluctuations of the wave into a

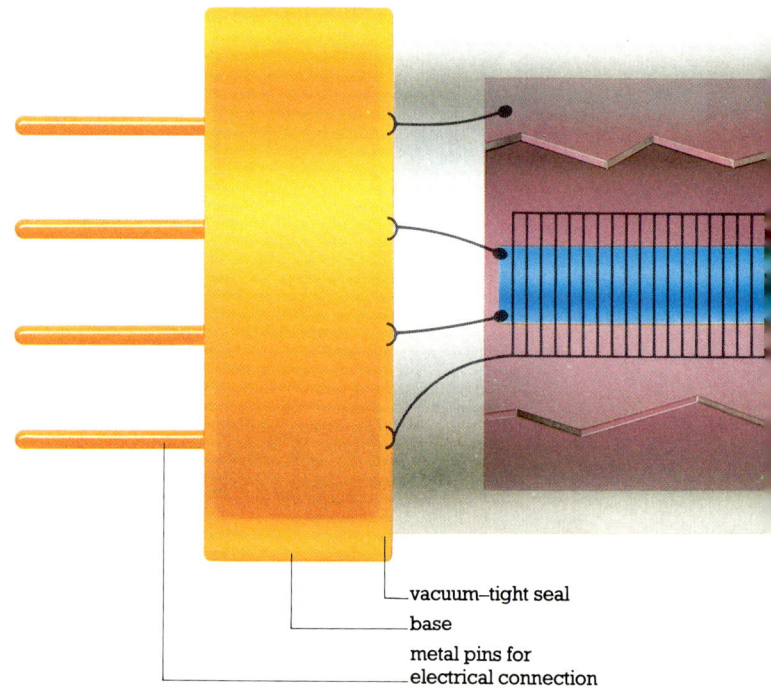

vacuum–tight seal
base
metal pins for
electrical connection

Above is a drawing of a triode. It has three electrodes positioned in an air-free tube. In the electrical diagram below, if an input signal is applied to the grid, it emerges from the anode very much amplified.

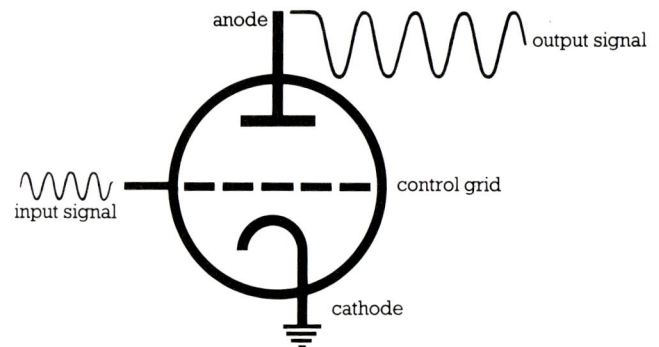

anode
output signal
input signal
control grid
cathode

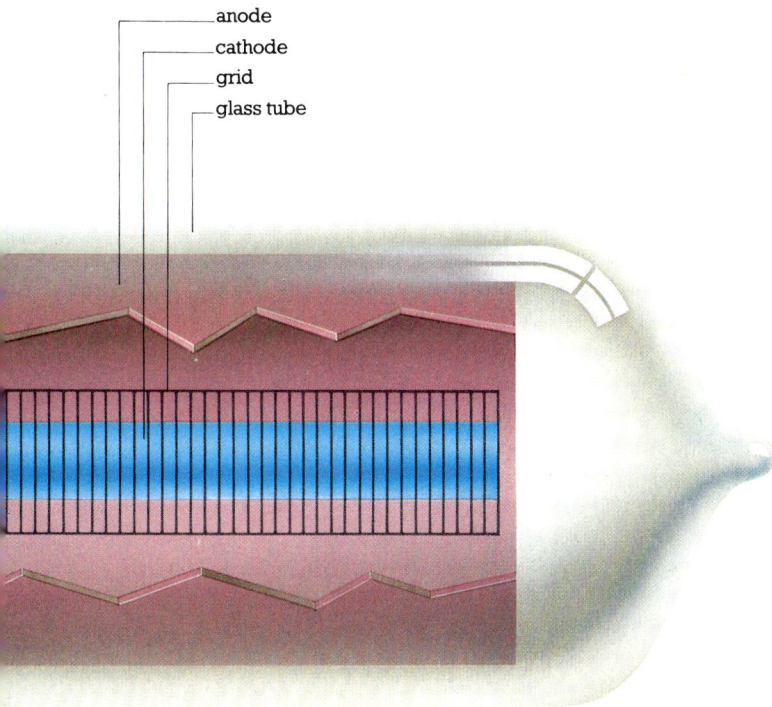

anode
cathode
grid
glass tube

Below is a valve, or vacuum tube diode, similar to the one devised by John Fleming in 1904. In a diode a heated filament releases a stream of electrons that flow towards a positively charged plate.

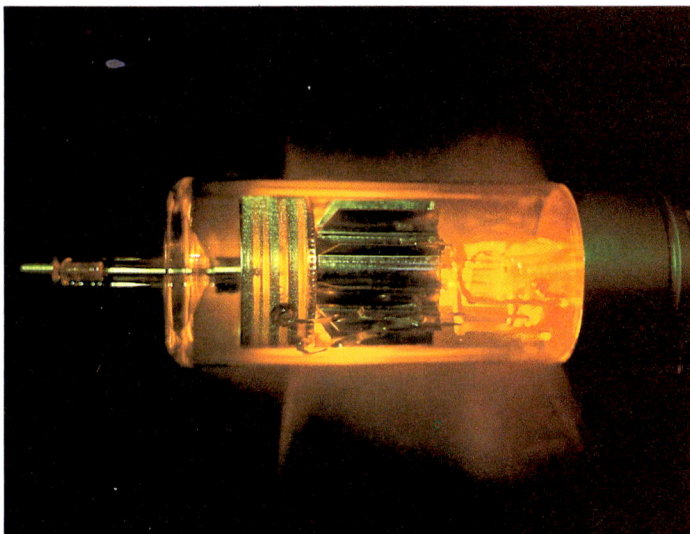

series of positive pulses that corresponded to the dots and dashes of the Morse code.

Because Fleming's tube let current flow in one direction but not in the other — just as the valve in a tire lets air flow in but not out — he called his invention a valve. Later, similar devices were called vacuum tube diodes, because they had two electrodes (*di* is Greek for two).

Later diodes were used to recover sound from radio waves. Unlike the dots and dashes of Morse code, which are transmitted as short or long bursts of radio waves, sounds are transmitted as changes, or modulations, in the strength of a continuous wave. If both the negative and positive halves of these modulated waves were somehow to pass to the speaker of a radio receiver, the halves would cancel each other out and the speaker would not produce sound. But when a diode passes only the positive half of the modulated wave through the circuit, the signal is able to make the speaker vibrate and reproduce the transmitted sound.

In 1906 the American engineer Lee De Forest invented a tube with a third electrode, called a grid, positioned so that the electrons passed through spaces in the grid on their way from the cathode to the anode. If the grid was positive, it would accelerate the flow of electrons towards the anode, increasing the current. If the grid was negative, it would repel the electrons, reducing the current. A small increase in the charge on the grid would create a large increase in the current flowing through the tube. This enabled the triode — as De Forest's three-electrode tube become known — to boost, or amplify, weak radio signals — a truly revolutionary development. With the triode, radios could communicate at greater distances than ever before, and, for the first time, radio transmission of speech and music was practical on a large scale.

After vacuum tubes were put into mass production during the 1920s, they became essential parts of all kinds of devices from radios and televisions to radar and transatlantic cables. Improvements in the manufacture of tubes extended tube life from a few weeks in 1915 to more than ten years in 1960. By this time, however, vacuum tubes were nearing the limit of their usefulness. Today, they have been replaced almost entirely except in special devices, such as high-power transmitters.

Electronics Today

By the 1940s it was clear that the days of vacuum tubes were numbered. They required too much power, produced too much heat, and were not reliable enough for the complex circuits being developed. Scientists began looking for something to replace tubes and the most promising line of research involved solid-state devices, in which current flows through a solid material rather than a vacuum. The materials of solid-state components — most often silicon — are called semiconductors because they can conduct electricity, but not as easily as true conductors such as copper and silver.

Ordinarily, semiconductors have very few free electrons, which are needed to carry a current. But if they are modified, or "doped", by adding less stable materials such as arsenic or phosphorus, they will contain more electrons than when they are pure. If they are doped with materials such as aluminum or boron they will contain fewer electrons, creating "holes" that will easily accept free electrons. Semiconductors with extra electrons are called n-type (n for negative); and those with extra holes are called p-type (p for positive).

If a bar of silicon is doped so that one end of it is p-type and the other n-type, the bar can perform a very useful function. When the n-type end is connected to the negative terminal of a battery, and the p-type to the positive, the extra electrons and holes will be repelled from either end and move towards the junction between the two. The electrons from the n-type end move easily across the p–n junction into the holes, creating a current. If the battery terminals are reversed, the electrons and holes will be attracted towards the terminals and away from the junction, cutting off the current. This simple semiconductor does the work of a vacuum tube diode: it allows current to flow in one direction, but not the other. The action of electrons and holes at the p–n junction is the fundamental principle of solid-state

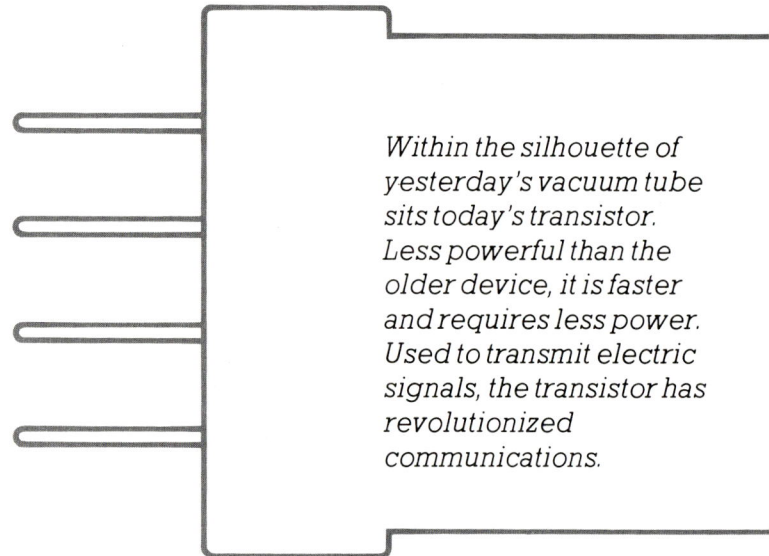

Within the silhouette of yesterday's vacuum tube sits today's transistor. Less powerful than the older device, it is faster and requires less power. Used to transmit electric signals, the transistor has revolutionized communications.

The transistor shown here consists of one piece of silicon modified with different impurities to give it three layers. The end layers have one kind of electrical property, p-type; the middle has the opposite kind, n-type.

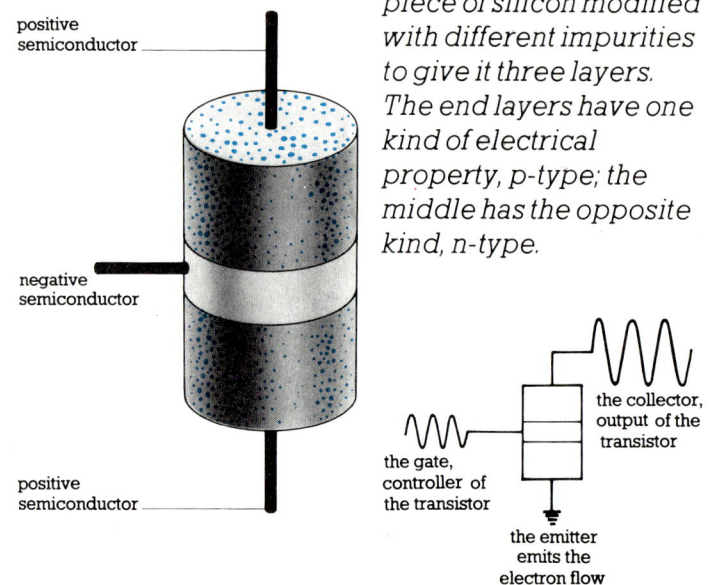

positive semiconductor

negative semiconductor

positive semiconductor

the collector, output of the transistor

the gate, controller of the transistor

the emitter emits the electron flow

electronics. A solid-state diode has one p–n junction; a transistor — roughly speaking, the solid-state equivalent of the triode — typically has two.

When the transistor was invented in 1948 it transformed electronics. At last a semiconductor device could do everything vacuum tubes did — and do it faster, more reliably, and without needing as much power. It could also be very small — the first transistors were no bigger than a vitamin capsule.

During the 1950s transistors began to replace vacuum tubes in common communications devices. The greatest impact of transistors, however, was on a relatively new invention: the digital computer. Because transistors could rapidly switch a current on and off, they were ideal for these new computers, in which an on or off current represents 1 or 0, the digits that are the basis of the computer's language. Just a few interconnected transistors make the elementary circuits that are the building blocks of highly complex computer systems.

In 1958 a remarkable invention, the integrated circuit, made transistorized circuits still smaller, faster, and cheaper by creating an entire circuit on a single wafer, or "chip", of silicon. In earlier circuits transistors, small as they were, were individual objects. On an integrated circuit, however, they are no longer visible as separate components. Instead, a piece of silicon is doped to create many microscopic n- and p-type regions within it in a prearranged pattern. This pattern determines which parts of the chip function as transistors, and which as other components. Microscopic metal channels connect the transistors and other components that have been formed on the chip to complete the circuits.

The technology which introduced integrated circuits is now referred to as Small Scale Integration (SSI). It contains fewer than 100 transistors on a half-inch square silicon chip. Large Scale Integration (LSI) — with more than 1,000 transistors per chip is now common, and the next phase, Very Large Scale Integration (VLSI), is now beginning. With VLSI, engineers are able to fit over 150,000 transistors onto a piece of silicon about a quarter of an inch square.

The incredible advances in digital computers made possible by today's electronics have spilled over into every area of communications. Digital electronics are now so convenient, inexpensive, and sophisticated that it is often more efficient to transmit many kinds of information — speech and pictures as well as computer data — in digital form.

The picture at left is of a silicon chip. The blue rectangle at center may contain 100,000 transistors or other electronic components. Tiny gold wires extend from the chip to external metal leads radiating from the central unit.

Black and White Television

The picture on a television screen is created by light that is constantly moving and changing its level of brilliance. Because these changes are very rapid, the human eye is unable to see each change and perceives them as uninterrupted movement. If a movie film is stopped, an entire picture will be seen, but if a television picture is frozen in time, it will theoretically show only part of a picture because the rapid changes will be caught in midstream.

While many inventors had been working on the idea of television before this century, it was not until 1925 that the English inventor John Logie Baird, and, in the United States, C. F. Jenkins, introduced the first mechanical television. In the Baird system, a camera lens formed an image of an object and focused it onto a light-sensitive electric cell. Between the camera lens and cell there was a rotating disk with a spiral of tiny holes in it. This moved across, or scanned, the image in lines to produce a flickering light, and thus a fluctuating electric signal. In a receiver this signal was passed to a light source located behind another rotating disk with a similar spiral of holes. As the light source changed in strength the lines traced by the holes in the scanning disk created a rough replica of the original object.

In 1928 the Russian-born American physicist V. K. Zworykin replaced Baird's mechanical system with an all-electronic one. Between 1930 and 1940 this system was further modified and improved and after World War II public television broadcasting developed rapidly to its present state.

All television is now electronic. A camera lens focuses an image of a scene on a picture tube whose light-sensitive surface develops a pattern of electric charges that represent the image. This surface is then scanned by a beam of electrons on a grid, or raster, of closely spaced lines. This happens in much the same way that you are reading the lines of this page right now. The electric charges on the light-sensitive surface act to alter the flow of the electrons, which in turn convert the visual information of the image into a varying electric current.

In the receiver there is a cathode ray tube (CRT) in which an electron beam scans a similar grid of lines

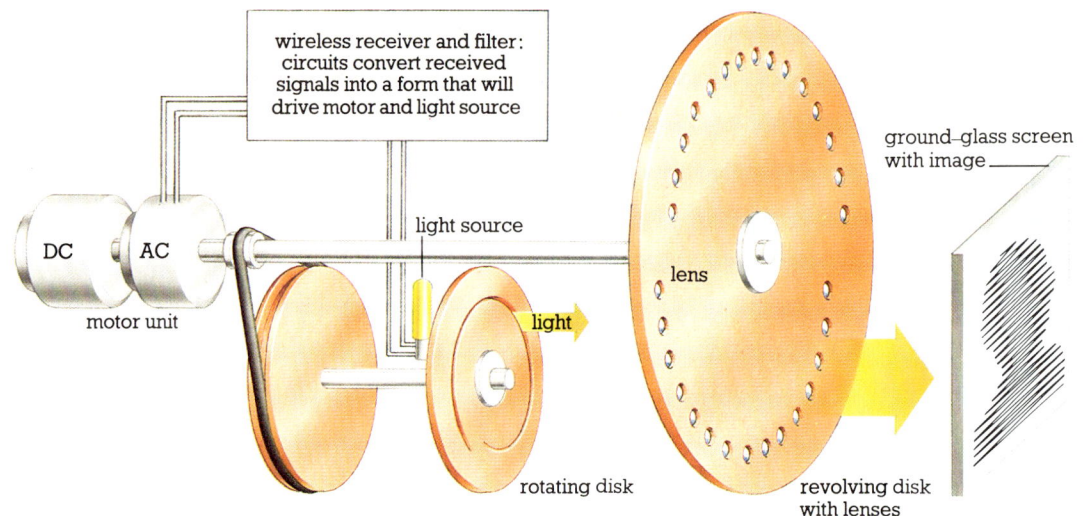

In the Baird mechanical television, a motor drove the device and a light source transformed electric signals into light. The light passed through a rotating disk with a spiral slot, onto a rotating disk with small lenses. This produced a flickering image on a ground-glass screen.

wireless receiver and filter: circuits convert received signals into a form that will drive motor and light source

DC   AC

motor unit

light source

light

lens

rotating disk

revolving disk with lenses

ground-glass screen with image

on a screen coated with chemicals called phosphors that light up when electrons hit them. The strength of the electron beam is varied, or modulated, by the changes in the strength of the signal received from the camera tube. Provided the two electron beams scan the grids in perfect harmony, an accurate replica of the original image is produced on the receiver's television screen.

Early television systems, in Britain for example, used a raster of 405 lines, scanning twenty-five complete grids per second. Today, to prevent flicker caused by the eye distinguishing one raster from the next, the rasters are interlaced. Each full raster, or frame, is made up of two half rasters, or fields, and the odd- and even-numbered lines are scanned alternately. In this way, sixty half-rasters, or fields, are scanned every second to produce a flicker-free picture of thirty full pictures, or frames, per second.

The sharpness, or "definition", of a television pic-

All black and white televisions use a scanning system. The definition of a picture is determined by the number of lines used for scanning. The United States, Canada, and Japan use a 525-line picture, and in Europe a 625-line picture is used.

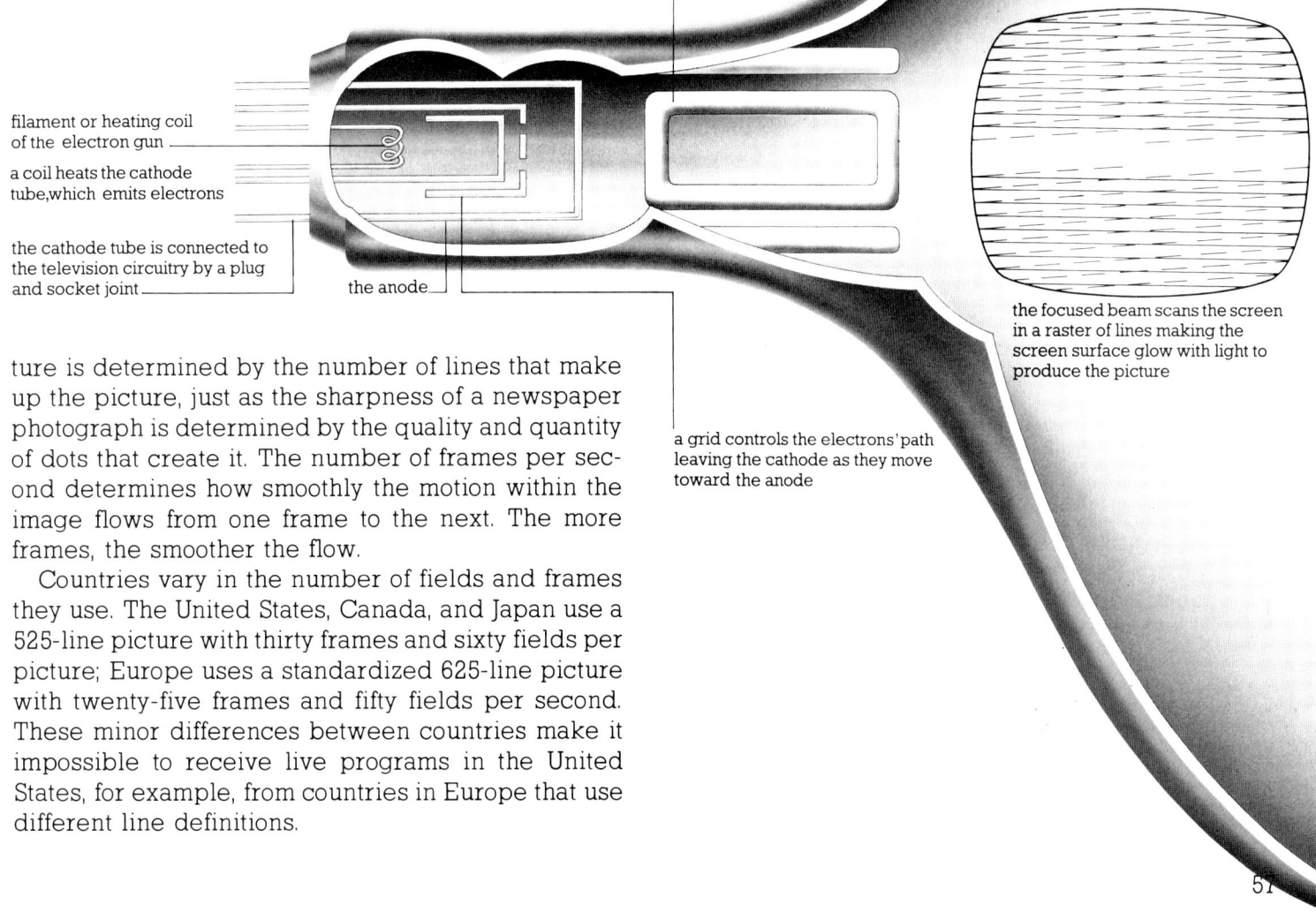

some electrons pass through the anode and are focused into a beam by electric coils which make the beam scan the screen in lines

filament or heating coil of the electron gun

a coil heats the cathode tube, which emits electrons

the cathode tube is connected to the television circuitry by a plug and socket joint

the anode

the focused beam scans the screen in a raster of lines making the screen surface glow with light to produce the picture

a grid controls the electrons' path leaving the cathode as they move toward the anode

ture is determined by the number of lines that make up the picture, just as the sharpness of a newspaper photograph is determined by the quality and quantity of dots that create it. The number of frames per second determines how smoothly the motion within the image flows from one frame to the next. The more frames, the smoother the flow.

Countries vary in the number of fields and frames they use. The United States, Canada, and Japan use a 525-line picture with thirty frames and sixty fields per picture; Europe uses a standardized 625-line picture with twenty-five frames and fifty fields per second. These minor differences between countries make it impossible to receive live programs in the United States, for example, from countries in Europe that use different line definitions.

Color Television

Light waves are different colors depending on their frequency. When red, green, and blue light are combined in the proper proportion they create what is known as white light. Creating colors by mixing light itself should not be confused with creating colors by, say, mixing paint which only reflects light.

Color television is based on the principle that different mixtures of red, green, and blue light can produce every color in nature. In the early 1950s engineers succeeded in putting color into television pictures by covering the camera screen with red, green, and blue filters in rapid succession, and scanning a separate field through each filter. Simultaneously, similarly colored filters moved across the screen of the receiver inside the set as the appropriate field was recreated. To avoid flicker, the filters and fields had to move very fast. The color picture obtained was good, but color broadcasts could not be reproduced on ordinary black and white televisions, and this system was never put into use.

Today, color television uses a picture tube with a screen surface peppered with clusters of tiny dots or lines that are composed of three different types of phosphors. When struck by a beam of electrons, some of the phosphors emit red light, some blue, and others green. Since the dot or line clusters are too small to be individually seen in normal television viewing, the picture produced on the screen appears to be a true representation of natural color.

The difficulty is to ensure that the electron beam strikes the right color phosphor at the right time. Engineers have tried to do this by guiding the electron beam so that it scanned the red, green, and blue phosphors in sequence. So-called beam-indexing tubes, however, are difficult to mass-produce because the electron beam must be designed to hit each phosphor with absolute accuracy, requiring precise and expensive equipment.

The color tubes first demonstrated in 1950 — and still the most common today — use what is known as a shadow mask. The surface of the television screen is peppered with triads of red, green and blue phosphor dots. Three electron guns in the neck of the television tube produce three separate electron

Inside this color television camera three separate light-sensitive electron tubes each produce an electric signal. One signifies the red content of the image, one the blue, and one the green.

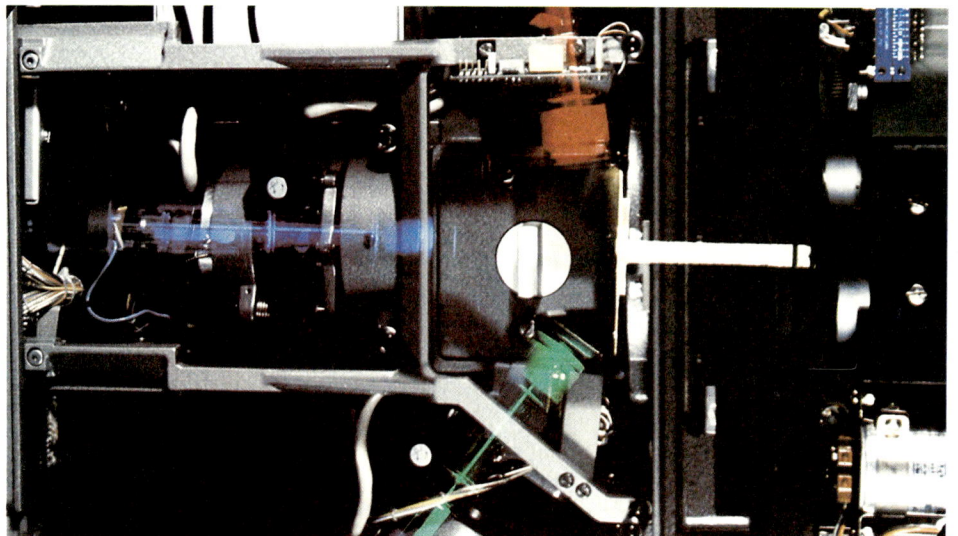

beams. A masking plate with a pattern of perforations that match the pattern of the phosphors lies over the screen surface, assuring that the beam from the red gun can only reach the red phosphors, the blue gun the blue phosphors, and so on.

In a modified version of this tube, called the Trinitron, the phosphors are arranged in vertical lines stretching from top to bottom of the screen. A grill with slits to match the phosphor lines overlays the screen, again ensuring that the red beam hits only the red phosphors, and so on.

In a modern color television, electron beams scan the screen surface in a grid of lines as in a black and white television. For color television, however, the strength of each beam is varied to produce a continually changing mixture of color balance and brilliance.

Black and white television broadcasting started long before color television and so each country had to adopt a color system compatible with its black and

In a modern color television, three electron beams scan a screen surface in a grid. The strength of each beam is varied to produce a changing mixture of color balance and brilliance.

the beams are focused and scanned by magnetic coils

three electron guns produce three separate beams of electrons

a color television connection is complex; the three electron guns must be fed with separate red, green, and blue signals

a shadow mask guides each electron beam to hit the correct phosphor

electron beams

unactivated phosphors remain gray in color

A

A: activated phosphors turn red, green, or blue

white system. This was so that the same television signal could produce color pictures on a color set and black and white pictures on a black and white set.

Compatibility between black and white and color transmissions is created by splitting the transmitted television signal in two. One half, the luminance signal, contains all the information about picture detail and brightness. In a black and white television set the luminance signal drives the single electron gun. The other half of the signal, the color, or chrominance signal, is carried piggyback on the luminance signal and contains all color information. In the receiver, special circuits decode the color signal and recombine it with the luminance signal. Three separate signals are then produced to activate the red, green, and blue electron guns of the color tube.

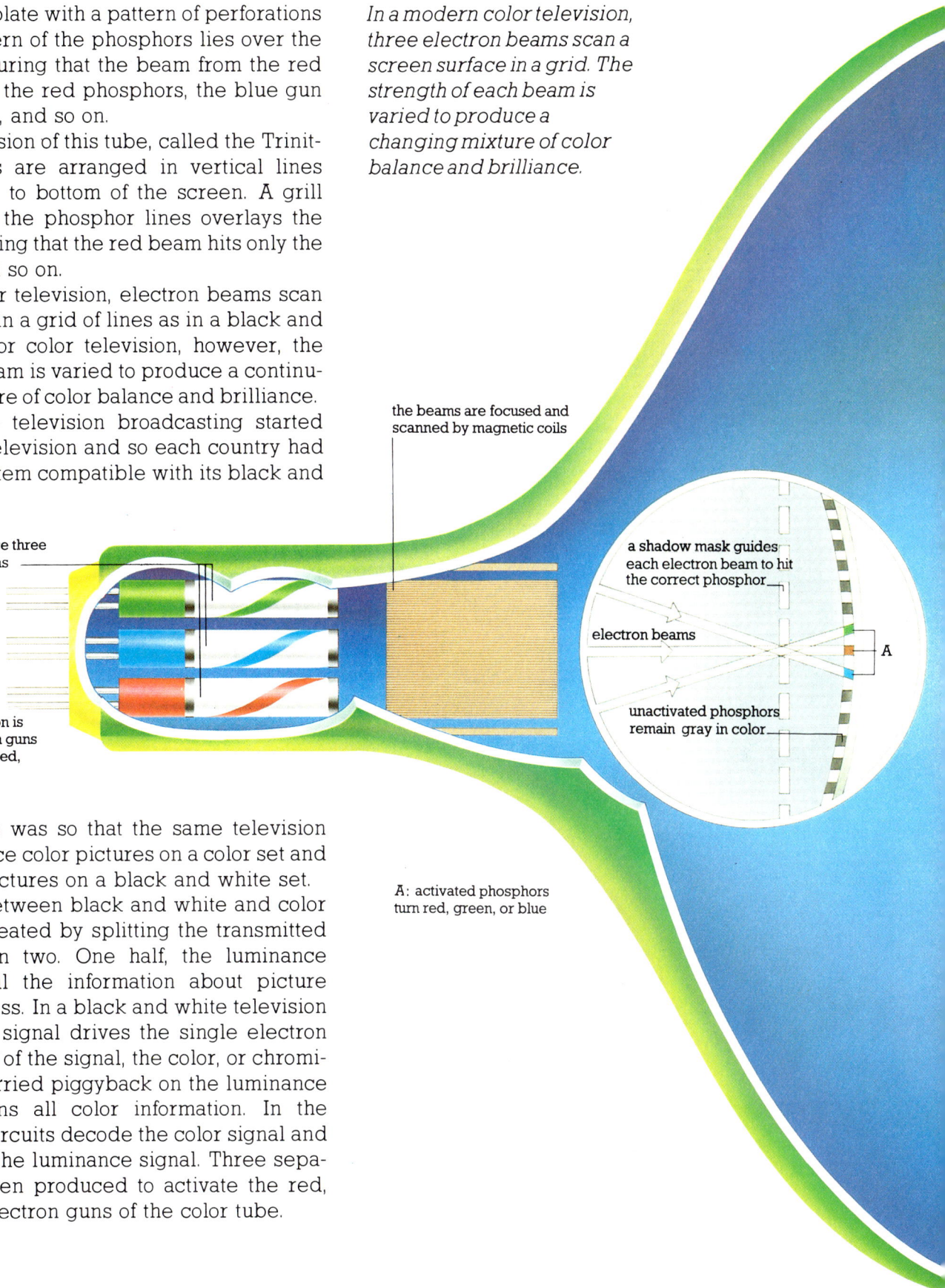

Flat Screen Television

A flat screen television, as the name implies, is one in which the display screen and all of its components make up a single relatively thin panel. While flat screen televisions are now available to the public, scientists have not yet been able to match completely the picture definition or low cost of a conventional televison set. At the moment, large flat screen televisions are used to display televised sporting or ceremonial events in public places.

In one version of a flat screen television thousands of colored electric light bulbs are arranged in clusters of red, green, and blue (the size of the bulbs depends on the size of the screen). The bulbs are activated by continually varying electric currents so that each cluster changes brilliance to produce a constantly changing mixture of colors. When viewed from a distance only a color picture is seen, and not each cluster of the individual bulbs.

In smaller flat screens now in use — mainly by the military — the electric light bulbs are replaced by points of light that are produced by tiny cells filled with gas or by light-emitting diodes, triggered by a changing electric current. The difficulty with this technique is that manufacturing light sources small enough to create a picture of high definition is very expensive. Similar to a newspaper photograph, the smaller the points of light, the clearer the picture. The size of the points is relative to the size of the screen so that the smaller the screen, the smaller the points will need to be.

An answer to this problem could be provided by liquid crystals of the type used in digital watches and calculators that show black numbers against a gray background. Liquid crystal requires little electric power because it uses ambient, or natural, light instead of having to produce its own. An electric signal changes the optical characteristics of the ordinarily translucent crystal, making it opaque. Early liquid crystal displays were unreliable, did not last long, and produced grayish pictures lacking good tonal contrast. Today, crystal cells have been vastly improved and can last for tens of thousands of hours. However, because liquid crystal does not generate

In the near future, flat screen televisions will be similar to this model. It uses a new type of screen that allows the housing of the set to be thin, small, and lightweight.

its own light it cannot easily be made to produce a color image — it will simply reflect the color of the light around it. Liquid crystal also responds slowly to changes in electric signals.

All flat screen televisions, whether they use gas-filled cells, spots of light-emitting material, or liquid crystals, require a complicated network of electric connections to trigger each picture point. It is impractical to provide each picture point with separate wires because even a small screen would then require many thousands of connections. To solve this problem, a grid of connections made from horizontal and vertical wires is used to create a mosaic of crossing points. This means that a large number of points can be individually triggered into action by signals passed by a small number of wires.

The latest developments in flat screen television involve mounting a liquid crystal display directly onto a large slice of silicon, like the type used to produce microcomputer chips. Engineers, however, have not yet been able to make a screen larger than

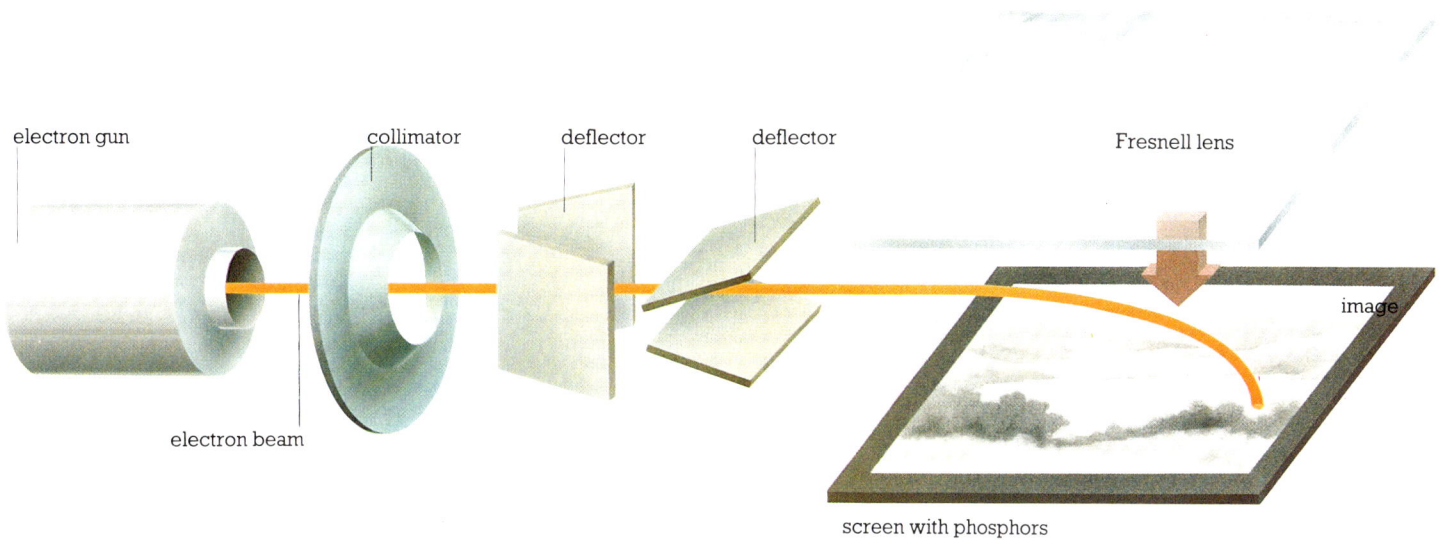

electron gun collimator deflector deflector Fresnell lens

electron beam

image

screen with phosphors

This diagram shows how one kind of flat screen television works. An electron gun produces an electron beam. This is focused and then deflected by plates so that it scans the viewing side of a phosphor-covered screen after bending through ninety degrees. The viewer watches the screen through a flat Fresnell lens that is used to increase the size of the small image.

forty-eight inches on a side. It is still also much more expensive to produce a flat screen television than a standard color television.

The cost and difficulty of making flat screen color televisions has encouraged engineers to look into the possibility of modifying conventional television picture tubes. Several manufacturers have already developed semi-flat television tubes that work much like conventional tubes but are flattened so that the electron gun lies alongside the screen firing its beam parallel with the screen surface. A series of magnetic coils then bend the beam through ninety degrees to scan the screen. Although no match for a real flat screen, tubes such as this are already being used in miniature television sets at a reasonable price.

The flat screen picture, right, is made from thousands of points of light. At present, the points are still visible in the image, but engineers are attempting to overcome this problem and match the picture quality of conventional televison sets.

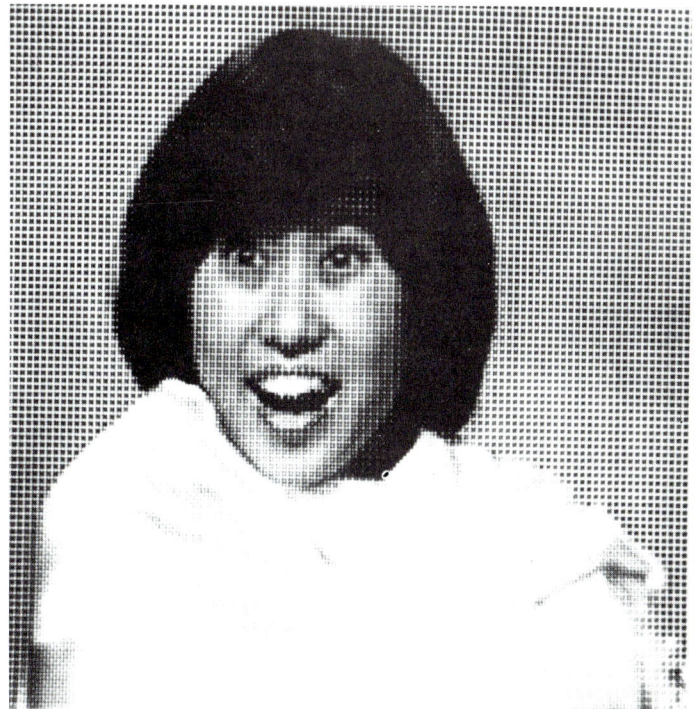

The Role of the Computer

We are now facing a revolution — the information revolution. Just as the Industrial Revolution changed people's lives throughout the world, so information technology will change our lives over the next few decades. Information technology is the marriage of communications and computers to supply information. This technology has created new uses for telecommunications that allow businesses and ordinary people easy access to vast amounts of information stored in computers. As engineers develop new information systems, they discover new things they can use them for.

In the telephone system, the modern Electronic Switching Systems (ESSs) that have largely replaced older electromechanical switching offices are actually computers themselves. They have memories and can store programs, as well as perform ordinary switching functions, such as sending dial tones or busy signals, and routing calls to the next switching office. They perform these functions so quickly that they have time in between to execute programs that test and monitor telephone equipment. In about the same time as it took for a call to go through in the previous systems, the ESS is able not only to make the desired connection, but also to check whether outgoing trunks are busy or idle, or whether telephone lines show that the phones are on or off the hook.

Digital switching centers can also provide new services to a telephone user such as transferring calls. If someone will be away from home for a day, for instance, his calls can be routed to the number where he will be staying. The switching center can store frequently called numbers, so that the caller dials only a short code instead of the whole number.

The revolution in information technology will save a lot of the time people now spend on trips to the office, school, and stores. As various systems link home computers with those in places of business, office workers will be able to do more of their work, students more of their studying, and families more of their daily business, without leaving the house. Soon, for instance, people may be able to shop from home. Using a terminal attached to their television set, they will call up a list of goods and prices that will appear on the television screen. They can then order the goods for delivery and pay by credit, using the same terminal. Already in some countries it is possible to reserve a theater seat or book a hotel room from your own living room.

In this picture a computer engineer is monitoring a printed circuit board. The information he is receiving can be fed into other computers for evaluation.

Though information technology has not yet had its full impact in the home, it has already begun to change life in the office. Even small businesses now use computers because of the need for more and easier access to essential data. They can use either fixed terminals in their offices or factories, with direct communication links with a computer, or portable computer terminals. Now there is even a hand-held computer terminal that can be connected to a computer over a normal telephone line. These terminals are used in a variety of jobs. A salesman, for instance, can check the stock available in his warehouse and record an order made by a new customer, right from the customer's premises.

Manufacturers are already mass producing small computers, with the result that the price of these new communications products is rapidly falling. In the future they will probably be as common in the home as the telephone is today.

In most airports, a complex computer system is used to continually update arrival and departure information.

The printer below receives words and pictures sent as digital signals from hundreds of miles away.

Viewdata and Teletext

An ordinary television set, with a special attachment and some modifications, can now be used to bring into a home the vast amounts of information stored in computers all over the country, and even to communicate with those computers. Two types of information systems are already in use — viewdata and teletext. Teletext is a passive system; it can only receive information. Viewdata, however, is active; the user can receive information from a computer and talk back to it. Both systems use a regular televison set that has been modified to receive a special signal.

With viewdata, an ordinary television set equipped with a computerized controller can receive information from a computer over telephone lines. Using the controller, a caller can search the memory of a distant computer and the information he desires will be displayed on the television screen. Thus, with viewdata

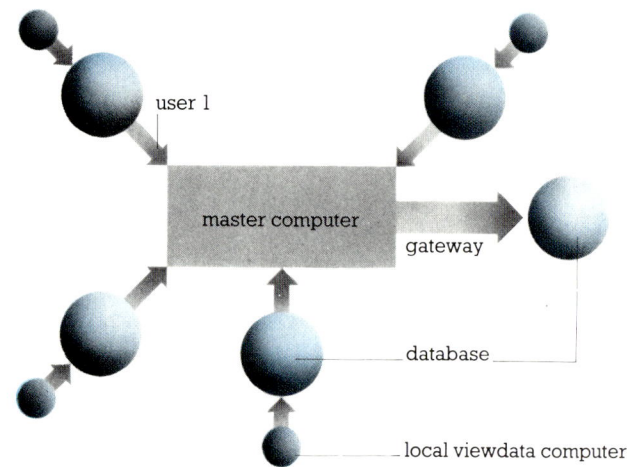

In the diagram above, a small local terminal gains access to a large database through a public viewdata network and the gateway system.

A travel agent helps a customer plan his trip with the assistance of viewdata. The terminal at the agent's right is connected over telephone lines to a central computer that stores useful data such as airline schedules and hotel vacancies. Once the customer has made his decisions, the agent keys in the information through the system and books the customer's entire trip at one time.

a private user has access to a large central computer which, because it serves many people using it for different purposes, holds a large memory bank with a wide range of information.

Teletext is a method of sending words and graphics over the air. The teletext signal rides piggyback on the signal that brings ordinary television programs into the home. The viewer uses a keyboard to choose the pages of information he desires but, unlike viewdata, the operator cannot talk back to the computer. The whole of the teletext database, or computerized library, is broadcast continuously. When a viewer tells the machine he wants a particular page, the computerized controller tunes in to display that page.

Television companies all over the world are experimenting with teletext, and services are now available in a number of European countries and the United States. Teletext can provide information on a variety of things, from weather and sports reports to news and entertainment guides.

Because viewdata allows two-way communication between people and computers, it has more applications than teletext, particularly for businesses. For this reason, in the next few years viewdata will probably attract more users in the office than in the home.

Viewdata, for example, can be used by travel agents to check seat availability on particular flights or trains. To do this, the agent calls up the viewdata service and asks for booking information. He can then, using a system called gateway, reserve a seat by typing the necessary details onto a keyboard. Gateway is the most important development in public viewdata. Before it was invented a user could only give simple responses to the viewdata services main computer, but with gateway he can use the public viewdata system to talk to a private computer. The travel agent, for example, goes through the viewdata network to an airline's computer, which answers his questions through a public system.

The potential uses of gateway are many. Already some banks are giving viewdata terminals to their customers on a trial basis so they can carry out their banking transactions from home. In the future you might go teleshopping from your armchair by ordering goods from local stores on credit and having them delivered as well—a particularly useful service for the elderly and disabled.

There are now many private viewdata systems

available. A very large company with vast stores of information might prefer a system of its own rather than sharing on a public system. With a private system a company uses its own computer to provide clients or employees with such information as prices and supplies, company accounts, or holiday dates. Public and private viewdata are also used for sending mail electronically. The message to be sent is typed on a screen and the number of the person to receive it is dialed just as on a telephone. The electronic message is stored in a "mailbox" in the computer, and the recipient is told there is a letter for him which he can either read immediately — or when he next goes to his television set.

The operator of a teletext machine normally uses a calculator-sized, hand-held remote control device to give the machine instructions.

Digital Communications: Voice

The men in this picture are transmitting messages to Westar II, 22,300 miles over the Equator. They are using an analog system with twelve channels. The analog voice signal is passed to a multiplexer where it is converted into digital signals and multiplexed. The digitized signal is passed to a modem and on to a dish antenna for transmission. Messages are beamed down by the satellite to another antenna in Sunnyvale, California where they are reconverted into analog signals.

Until recently most telephone circuits carried analog signals but because digital techniques are faster, have a greater capacity and are cheaper, in the 1960s they began to replace analog systems. Digital techniques were first used for short distance communications, on lines linking local telephone switching centers. With the advent of digital telephone switching, the extension of digital techniques to the entire telephone network was the next logical step toward improved telephone communications.

In the digital system, a telephone first converts the voice of a caller into analog signals, or electrical waves. These analog signals are then converted into digital pulses, or bits, by a technique known as pulse code modulation, or PCM. The strength of the analog signal is first tested, or sampled, 8000 times per second. The value obtained for each signal is then trans-

lated into a code made up of a combination of on and off pulses that represent the binary digits one and zero. Different combinations of eight digital pulses are sufficient to differentiate one sample from another. Mathematically this means that each digital telephone conversation is made up of a stream of pulses, 64,000 of which are transmitted each second.

These digital pulses are then transmitted to their destination, where they are converted back into analog signals. They must be turned back into analog signals because on a telephone earpiece the digital signal itself would produce only a hiss. This coding and decoding process is at present performed at the local telephone switching center, but in the future may be accomplished by the telephone itself.

There are considerable advantages with digital telephony. All telephone circuits suffer from

This is a console telephone, part of a so-called PABX system that switches both voice and computer signals throughout an office telephone network. A digital system, it can carry up to 12,000 internal lines and 4600 outside lines. A digital switching and transmission network offers greatly reduced noise and distortion and the simplicity of one network to handle both voice and data simultaneously.

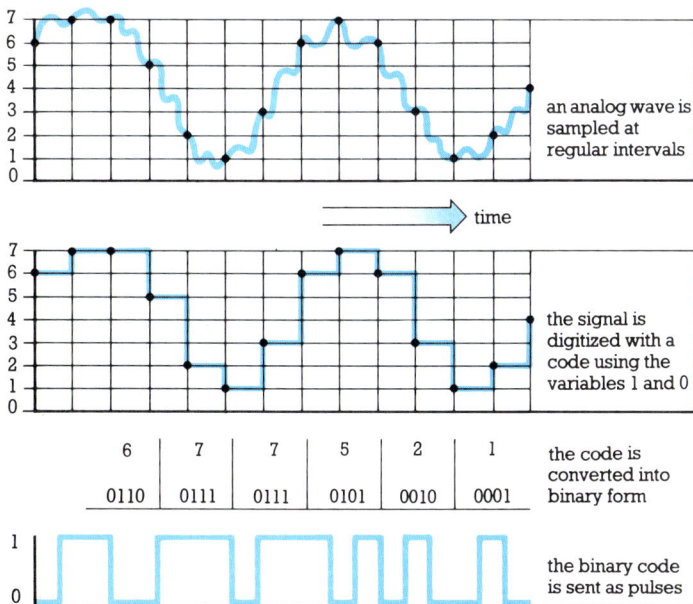

noise — interference caused by lightning or the electricity in nearby cables, for example. In the analog system, in which the message is carried on a continuously varying electric current, the noise can seriously distort a weak signal. This problem is virtually eliminated in a digital system because there is either a pulse of current or there is not. Thus such a signal can be discerned accurately through noise that would distort an analog signal.

All telephone systems require amplifiers or repeaters placed at regular intervals along the telephone line. These are used to boost the strength of the telephone signal, which weakens with distance. In an analog system these repeaters simply increase the strength of whatever signal is received — including unwanted noise. In a digital system each repeater generates a new set of pulses. No noise is sent on.

A digital telephone system has a number of economic benefits. Firstly, it can be less expensive to construct than an analog system. Silicon chip products are generally smaller, more versatile and less expensive than traditional telecommunications products and are better suited to digital systems. Secondly, digital signals can be transmitted over fiber optic cables, while analog signals cannot.

Lastly, a completely digital telephone system will increase the options available to the user. Digital signals, for instance, are easier to store than analog. If the desired telephone number is busy or unavailable, the caller can dictate his message into the telephone to be stored and the system will deliver the message at a time when the number becomes available. Other new options include picture or videophone, supertelex, facsimile and viewdata services for the householder — all sent down the same single set of wires.

an analog wave is sampled at regular intervals

time

the signal is digitized with a code using the variables 1 and 0

6	7	7	5	2	1
0110	0111	0111	0101	0010	0001

the code is converted into binary form

the binary code is sent as pulses

Many communications services are now using digital means to transmit voice signals because the signals can be grouped and passed far more efficiently than by using analog signals. The diagram at left shows how analog signals are converted into digital signals. The analog wave is sampled thousands of times per second. Each point sampled is assigned numerical value and then converted into binary form, which expresses all values as some combination of the digits one and zero.

Digital Communications: Data

Modern computers first came into service in the late 1940s and 1950s. Known as number crunchers, these were designed to do long, complicated arithmetic calculations very quickly. The calculations made by the computer were printed out, and carried or mailed to those needing them, a very time-consuming task.

As the computer became more powerful and sophisticated, it was used for more important tasks. Eventually the mail proved too slow for delivering urgent computerized information, for example that needed by air traffic controllers. To quicken the process, the computer was connected to the telephone. In many ways this gave economic advantages. For example, instead of having a large, expensive computer at each site where it was needed, a simple terminal with a keyboard could be installed at the workplace. Many terminals could be connected to a large central computer by telephone so that a large number of different users could call on the computer.

Telephone lines, however, are not really suitable for handling computer traffic because they are designed to carry analog signals whereas computers speak in digital languages. Computers can also communicate hundreds of times faster than people; so fast, in fact, that a computer can send all the words in this book to another machine 100 times every second.

To send digital computer data on an analog voice circuit it must first be disguised as an analog signal. Engineers call the electronic box that does this a modem, a contraction of the phrase *MO*dulator-*DEM*odulator. Modems work in a number of ways, but their main function is to convert digital output from a computer terminal into analog signals. When the signals arrive at their destination, another modem converts them back into the digital form required by the computer to process them.

Even this system suffers from a number of disadvantages. Firstly, computers and terminals can only communicate at a fraction of the speed at which they

work. Secondly, ordinary telephone lines from time to time suffer from interference and noise that can so badly distort the disguised computer data that the receiving modem is sometimes slightly confused. For these reasons, communications authorities have developed networks dedicated solely to computers. There are two main types of computer networks—circuit-switched and packet-switched.

With circuit-switching, once a connection is made between two computers or their terminals, no other

At an airport communications center, digital computers are used to identify the address codes of incoming and outgoing planes, and switch each message to an agreed routing plan.

machine can use that line. The link between the computers or terminals is maintained until all the data has been transmitted.

Packet-switching is more efficient because it allows the network to be shared by all users at all times. This is possible because the data is split into packets, each carrying the address of its destination in a computerized code. The packets move through the network alongside packets from other computers going to different places. At their destination, switching machines read the coded addresses on each packet and deliver it to the correct computer.

There are now networks for specialists in various fields that allow people operating computer terminals to ask all over the world for scientific, technical, medical, and economic information. These sophisticated systems work alongside a new generation of number crunchers, which are still performing their gargantuan duties.

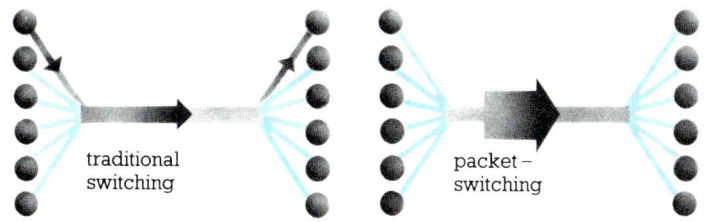

traditional switching

packet–switching

In the diagram above, a data link (left) uses a traditional switching technique in which a complete circuit is made between the sender and receiver. With this method no other sender can use the line until the first has completed its transmission. In packet-switching (right) data from all the senders is formed into packets, each with a different address, and sent down the line together. At the end of the line the packets are sorted and sent to their separate destinations.

This airport control center keeps track of all aircraft traffic on the runways by continually receiving and processing digital data.

Recording Sound: Records

Seen at right, greatly magnified, are the grooves of a stereo record. The grooves contain information for two separate sound channels. The walls of the grooves undulate in different patterns, creating different signals for either channel.

In the grooves of a 78rpm record (far right), the walls undulate together, indicating that only one sound channel is recorded.

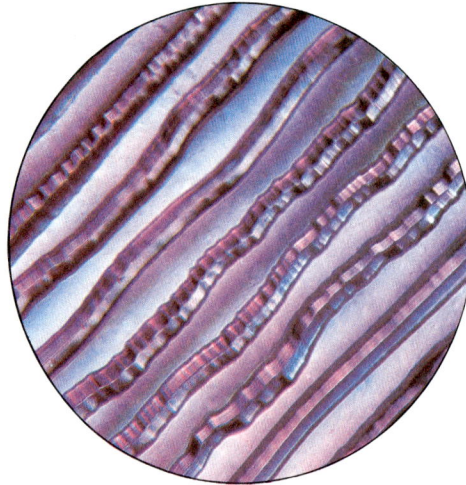

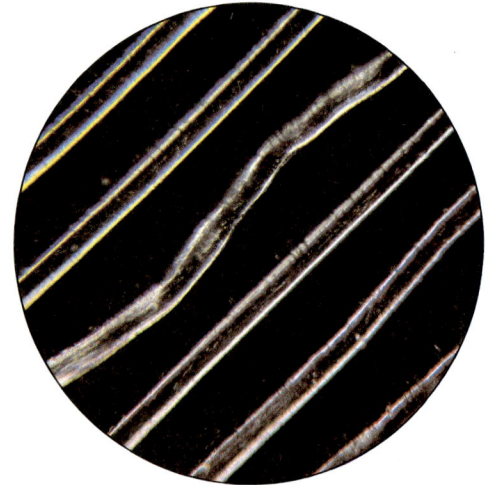

Sound travels as waves of air pressure. The higher the frequency, or the number of waves per second, the higher the pitch of the sound. The higher the pressure of the wave, the louder the sound. We are able to hear sounds because our eardrums vibrate in sympathy with these waves of air pressure. To record sound, pressure waves (sound waves) must be captured and frozen in some way.

Records are the oldest means of capturing sound in use today. They trace their heritage back to 1857, when Léon Scott de Martinville, a Frenchman, succeeded in visually recording sound. The de Martinville phonautograph, as it was called, traced a representation of pressure waves onto a cylinder covered with lamp black, a substance made from soot.

The real breakthrough, however, came twenty years later with Thomas Edison's invention of the phonograph. Now sound could not only be recorded, it could also be played back. In place of de Martinville's lamp black, the cylinder in Edison's phonograph was coated with tinfoil. A diaphragm, vibrated by the sound waves, moved a pointed needle, or stylus, that in turn cut a spiraling groove similar to the thread of a screw into the tinfoil. How deeply the groove penetrated the foil varied according to the pitch and strength of the sound waves received by the diaphragm. This method of recording was known as a hill-and-dale, or vertical cut, recording. To replay the recording, a lighter diaphragm and stylus were used to follow the groove already etched onto the cylinder during the recording process.

Ten years later in 1887 the German scientist Emile Berliner made the first gramophone. In this device, instead of a cylinder with an undulating groove spiraling from end to end, Berliner used a flat disk with a spiral groove that undulated from side to side. This was known as a lateral cut recording. The main advantage of the Berliner disk was that it was easy to duplicate in large quantities. A negative plate of the original recording could be used to make many identical copies. This could not be done with the cylinder technique.

Early records were ten or twelve inches across, made of shellac, and moved at a standardized speed of seventy-eight revolutions per minute (rpm). These first records were fragile and are today regarded as curiosities and antiques. After World War II, the long-playing, or LP, record was developed and today

In an elaborate recording studio (left) technicians stand at a mixing desk preparing a recording for stereo. Musicians often record a number of separate tracks using different instruments. The tracks can then be mixed together.

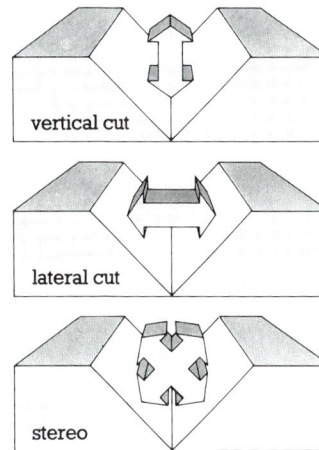

Diagrammed at left are three types of record grooves. In a vertical cut the stylus moves up and down. In a lateral cut, the stylus moves from side to side, and in stereo the stylus moves in both directions at once.

is still the world standard. A twelve-inch LP rotates at thirty-three and one-third rpm, has a more precise groove than a seventy-eight rpm shellac record, and can hold up to thirty minutes of sound on each side. This is about six times the capacity of the earlier shellac record. The seven-inch, forty-five rpm record is the other type now available. Originally developed as a competitor to the LP, it became the world standard for short recordings of popular songs known today as singles. Although once in bitter competition, the LP and single now peacefully coexist because most record players can play both.

A new grooveless record may soon surpass both the LP and single. Instead of grooves, a coating of reflective material covers a spiral track of microscopic pits. The pits represent sound converted into digital codes, which can be read by a laser beam. As the record rotates through the laser beam, the pits rapidly alter the reflection of the beam. These rapid alterations are read by a light sensor. The sensor converts the flickering light into electric signals which are reproduced as sound. Because light is used instead of a stylus, these digital records will not wear out like today's LPs and singles.

This is Edison's phonograph, the first device able to record and replay sound. To record, the phonograph etched a cylinder with a pointed needle that moved with varying sound waves.

71

Recording Sound: Tape

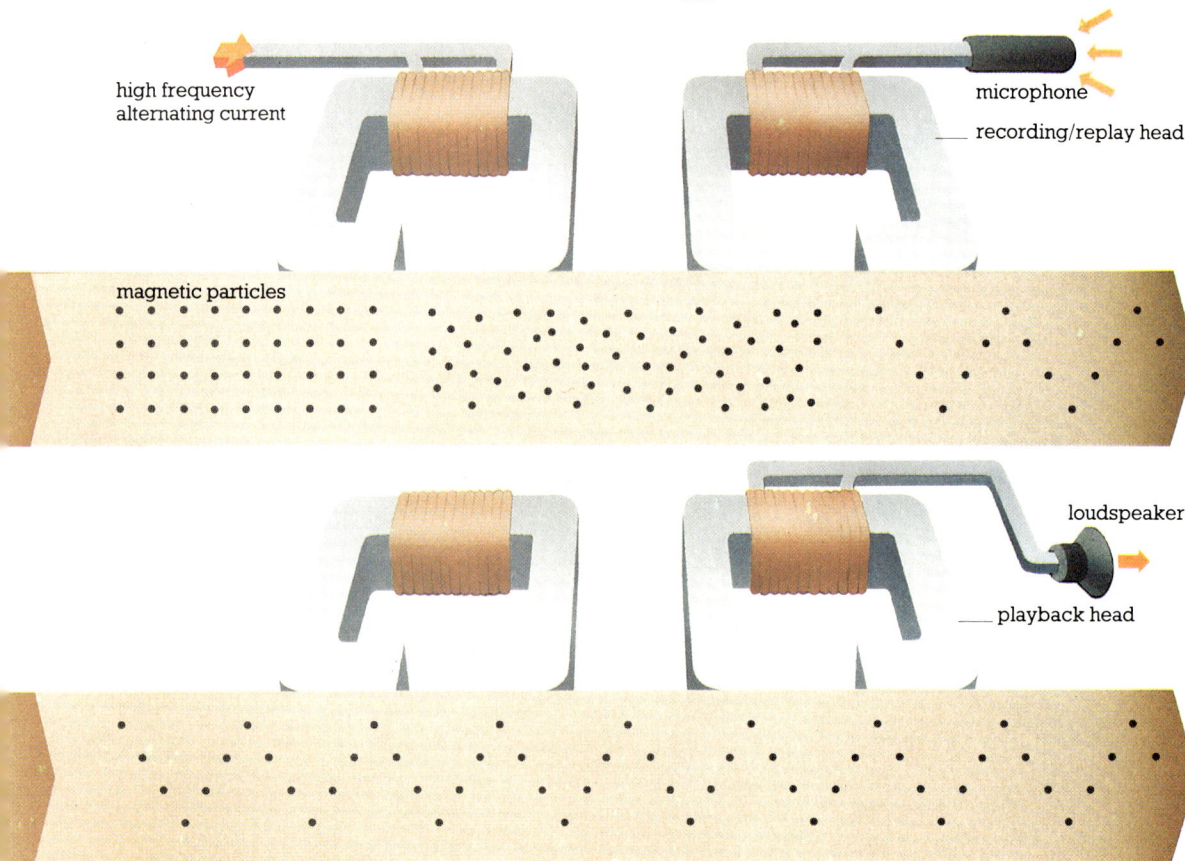

high frequency
alternating current

magnetic particles

microphone

recording/replay head

This diagram shows how sound is recorded on tape. The tape is coated with magnetized material. Electric signals are fed into an electric coil to create a magnetic field. As the tape passes by the head, the magnetic coating of the tape is altered by the changing signals.

loudspeaker

playback head

Here, a recorded tape is being replayed. The tape is run past a playback head. The magnetic pattern on the tape generates an electric current in the head's coil. The current is amplified and passed to a loudspeaker where it is converted into sound.

During the 1870s and 1880s various inventors in Europe and the United States suggested the idea of magnetic recording, and in 1898 the Danish inventor Valdemar Poulsen patented a working machine. Poulsen's machine, however, made recordings on steel wires, and it was not until 1934 that two German companies jointly produced the first recording machine using magnetic tape. Work continued in Germany during World War II and in the United States following the war. The enthusiasm for magnetic recordings expressed by manufacturers in the United States was due in part to the American singer Bing Crosby who quickly recognized its potential.

Magnetic tape is an efficient and economic way of recording sound because it is reusable, and the sound can be edited mechanically, by cutting and joining the tape, or electronically, using two or three recorders. The most common tape today is a strip of plastic coated with magnetic material that usually contains an oxide of metal such as iron or chromium.

In making a recording, a microphone converts sound waves into electric signals and the signals are fed into the electric coil of the recording head. The current in the coil creates a fluctuating magnetic field. As the tape passes close by the recording head, the tape's coating is magnetically impressed by this continually changing field, thus recording it. To replay the recording, the tape is run past the play-

In this professional recorder computer microchips control the recorder's operation. Such functions as recording and playback are automatically adjusted for different characteristics of different kinds of tape.

back head at the same speed used for recording. As the tape passes, the magnetic patterns on the tape induce a fluctuating electric current in the head's coil. When amplified and used to power a loudspeaker, this current creates an accurate replica of the sounds originally picked up by the microphone.

To recreate a sound successfully is not easy. For example, the individual magnetic particles on the tape coating can produce a disturbing hiss, which occurs if the tape is running at low speed and the particles on the coating are large; small particles are necessary to capture high frequencies. The frequency range of the tape recorder also depends on the tape speed, because higher frequency sounds require longer stretches of the tape to be faithfully recorded and reproduced. At higher speeds longer stretches of tape pass the heads.

A tape recorder needs a bias signal, which is an inaudible, very-high-frequency tone superimposed on the sound being recorded. The bias signal is necessary to overcome what is called the magnetic

inertia of the metal coating. Magnetic inertia means that, just as it is easier to spin a heavy wheel faster once it is already spinning, it is easier to increase the strength of magnetism in a material once the magnetism has started to build. A recording made without a bias signal would sound distorted.

Until the 1960s almost all tapes were one-quarter inch wide. To provide good quality sound with minimal background hiss, the tape ran at either seven and one-half inches per second or fifteen inches per second. Since then the audio cassette, using tape one-seventh of an inch wide running at one and seven-eighths inches per second, has become increasingly popular. Audio cassettes are convenient because the user does not have to take the time to thread the tape past the heads. Because of continual improvements in both recorder design and tape technology the quality of some cassette tapes is now as good as the traditional one-quarter inch tape, which runs at much higher speed. In the future, modified tape decks will be able to record sound in digital code.

Traditional Hi-Fi

The purpose of high fidelity, or hi-fi, is thought by some people to be to give the most accurate reproduction of original sound. In theory, a hi-fi system can reproduce a nearly exact replica of the sound heard in, say, a concert hall or recording studio. In practice this is not really true because very few recorded performances are left in their original state. Modern recording techniques enable artists to build up a recording layer by layer, erasing mistakes as they go. One or two musicians, for example, can produce an unrealistically loud sound simply by amplifying the original music. People also argue that the aim of hi-fi is to reproduce a live musical performance as accurately as possible. But most live performances, if recorded this faithfully, would reveal disturbing flaws. The goal of today's hi-fi is therefore best defined as the accurate, audible reproduction of an electric signal retrieved from a recording or a broadcast transmission.

A hi-fi system is like a chain: its total strength is only as great as that of its weakest link. There is no point, for example, in buying an expensive amplifier if it is to be used with an inferior tape deck, or loudspeaker. The best radio or television will produce poor results if connected to an inadequate antenna; the same holds true of a hi-fi system.

Music is a combination of ever-changing frequency patterns that react with each other in complex ways. The human ear and brain are able to recognize these patterns even if the music is masked by background noise. So little, however, is known about the way we hear and the audio performance of electronic circuits that two hi-fi systems may appear identical on paper but actually sound quite different. For this reason engineers are trying continually to devise new techniques to correlate the sound that a hi-fi produces with what it is supposed to do in theory.

The heart of a hi-fi system is the amplifier, which takes in electric signals from the record, tape, or radio and boosts them to a level sufficiently powerful to activate one or more loudspeakers. Some people believe that the ideal amplifier is one that does no more than make the input signals bigger. Others argue that because most loudspeakers have their own electric characteristics, an amplifier should tailor the signal that it receives to drive a loudspeaker consistently.

A turntable must spin a record at a constant speed. If it does not, the sound reproduced will be slow or fast wavering and the music will sound distorted. The turntable should be acoustically isolated, that is, placed physically away from the amplifier and loudspeakers. If not, it will act like a microphone and pick up sound coming from the loudspeakers, send it back through the amplifier to the speakers and pick it up again, creating what is known as feedback loop. The sound will go around the system in circles, getting louder and louder until it produces a howl. Also, the arm that holds the stylus must be free of resonances or it will vibrate in sympathy with some musical notes and not with others.

Loudspeakers are often said to be the weakest link in the hi-fi chain, even though engineers have dramatically improved their quality in the last few years. A loudspeaker produces sound by creating sound waves with a cone or diaphragm. Low-frequency waves are produced by movements at a slow speed back and forth over a greater distance. High-frequency waves are produced by small, rapid vibrations. Low frequencies are therefore best produced by a large cone or diaphragm and high frequencies by a smaller cone or diaphragm.

The problem that faces all loudspeaker designers is that their product must sound good in any room, despite its acoustics. However, the sound produced by loudspeakers will vary depending on where they are used. For example, loudspeakers placed on an uncovered floor tend to produce deeper bass tones, and in rooms with high ceilings and little furniture the volume will not need to be turned as high as in a low-ceilinged, crowded room. Consequently a person's favorite recording may well sound disappointing or altered in an environment different from where it is normally played. Thus most hi-fi systems are equipped with tonal controls to enable the owner to adjust the sound.

In a recording studio (left) a musician accompanies prerecorded music with an electronic flute which is recorded separately and the tracks mixed together.

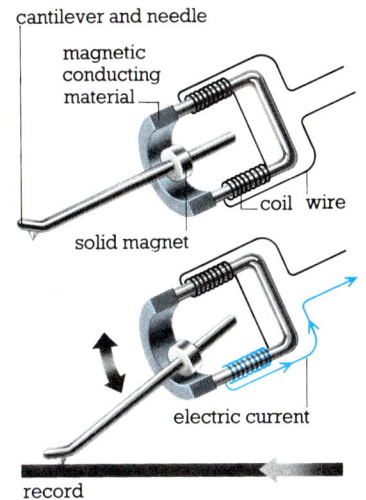

cantilever and needle

magnetic conducting material

solid magnet

coil   wire

electric current

record

Shown in cross section (above), one type of phono cartridge has a moveable magnet. The movements of the needle following the grooves of the record are carried by the arm to the magnet, which moves up and down and side to side within the ring of conducting material. A changing magnetic field is created and the coil transforms it into an electric current, which an amplifier will boost and send to the speakers, to then be turned into sound waves.

75

Stereo and Surround Sound

Two eyes enable people to judge depth and distance; two ears enable them to do the same with sound, as well as locate its direction. By using their eyes and ears together people can perceive a significant part of what happens around them.

A movie screen or photograph gives both our eyes the same view, resulting in a two-dimensional, flat image. Similarly, a single loudspeaker can only offer an unnatural, flat sound—like hearing an entire orchestra through a small window. An audience in a concert hall hears sound from all directions even though the orchestra may be grouped in a small area of the room. This is because the music produced bounces off the roof, walls, and floor many times, reaching the listener's ear as a complicated mixture of echoes. Unless the hall is extremely large so that the echoes take a long time to travel, the human ears and brain will not distinguish the individual echoes. Instead, a mixture of short echoes will be heard, which is called ambience, or acoustic warmth.

Early attempts to capture ambience and reproduce a sense of sound direction and depth date back to Alexander Graham Bell and early experiments to improve the hearing of the deaf. Bell pointed out that

Until recently, surround systems produced poor results. Future hi-fi systems, however, will create the illusion of surround sound much more convincingly.

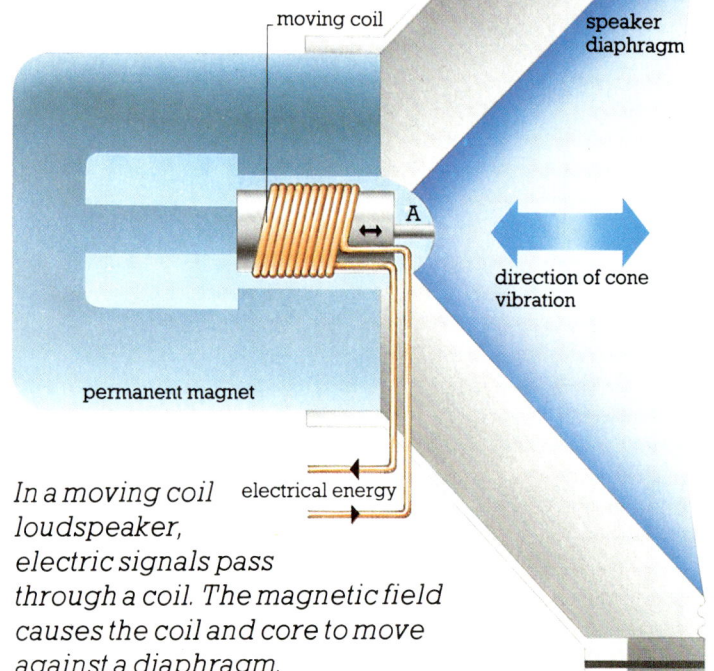

moving coil

speaker diaphragm

direction of cone vibration

permanent magnet

electrical energy

A

In a moving coil loudspeaker, electric signals pass through a coil. The magnetic field causes the coil and core to move against a diaphragm.

humans are binaural (in Latin, *bi* is two; *auris*, ear) and methods of sound reproduction would have to take this into account. A technique called binaural, or dummy head reproduction, does this.

When recording binaural sound an imitation human head has a recording microphone placed in each ear. (A dummy is used simply for convenience, to allow the person making the recording to move about during the process.) The recorded music is replayed through headphones so that a listener hears exactly what the microphones in the dummy's ear heard. In a binaural system each ear hears only the sound from one earpiece of the headphones, which creates a sensation of being surrounded by sound. Since 1930 engineers have tried to create this effect with loudspeakers.

The technique known as stereo captures sound through at least two microphones that point in different directions or are spaced apart so that they pick up different perspectives of the sound. These sounds are then reproduced after transmission or recording by a pair of loudspeakers that are spaced apart, pointing at the listener. Music coming from the left speaker is heard first by the left ear and then, a fraction of a second later, by the right. Similarly, sounds from the right speaker reach the right ear first and then the left. A pair of stereo speakers produces the illusion of a wall of sound, but unlike binaural, stereo cannot create the illusion that the listener is surrounded by sound as if in a concert hall.

Another approach, called the surround system, simply adds more loudspeakers, behind and to the sides of the listener. In the 1970s several firms offered surround systems known as quadraphonics, so-called because the technique used four loudspeakers. At least four different systems were developed, but none seemed to please the public.

The latest surround systems use four or more loudspeakers and process signals electronically. In this way an illusion of depth and ambience is maintained even when the listener moves away from the ideal listening position in the center of the room. The new surround systems work better than the old quadraphonic systems for this reason, but even these have not been a commercial success. Perhaps this is because record companies and the record-buying public have negative feelings left over from the old quadraphonic system. Ironically, the first stereo experiments in the early 1930s fared just as badly; stereo did not become popular until thirty years after the development of the first system.

A loudspeaker (left) is being tested in a specially designed chamber. The walls of the room are lined with sound-absorbent material to eliminate reflection and distortion.

At right, the different types of speakers available today. Omnidirectional speakers send sound in all directions. An electrostatic speaker has a diaphragm between two electrically charged plates. A standard speaker beams sound in one direction only.

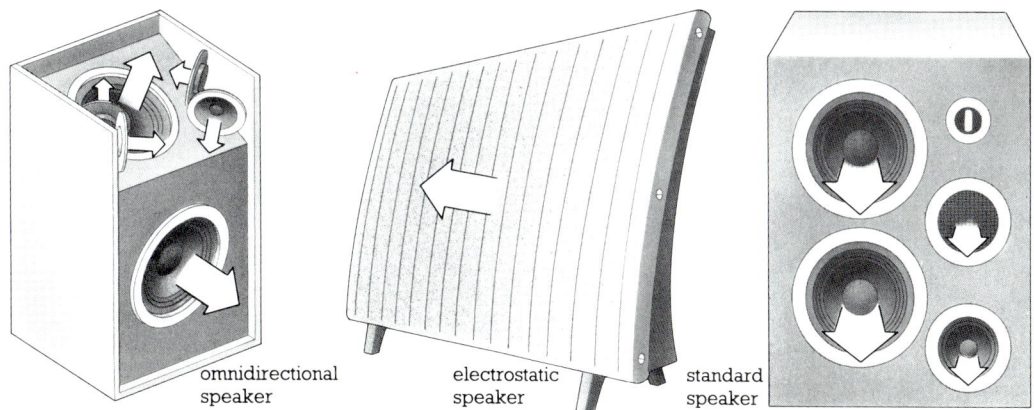

omnidirectional speaker

electrostatic speaker

standard speaker

Digital Hi-Fi

Most conventional records and tapes capture sound in analog form, which means that the undulations of the grooves on a record, or the changing pattern of magnetism on a tape coating, recreate — or are an analogy of — the original recorded sound waves. Now, however, audio engineers are beginning to use a different method of recording sound by converting analog waves into digital code. The digital pulses are recorded onto a record or tape and, when the record or tape is replayed, the digital code is reconverted into analog signals to produce sound waves. This works in much the same way as telephones that are now using digital signals rather than analog signals to transmit information.

The idea of converting analog information into a digital code is as old as finger counting. Electrical coding of analog information dates back to around 1840 when Samuel Morse invented his Morse code. A message sent in Morse code, however, relies on a human operator to change the words — or analog message — into a digital code. Nearly a hundred years later, in 1937, Alec Reeves, a British engineer working in France, designed a switching device that

In this recording studio a digital unit is used to mix sound. With such equipment three separate tracks of sound can be recorded and mixed simultaneously.

In the domestic digital tape recorder below the mechanics and circuitry of a video recorder are applied to the digital taping of sound.

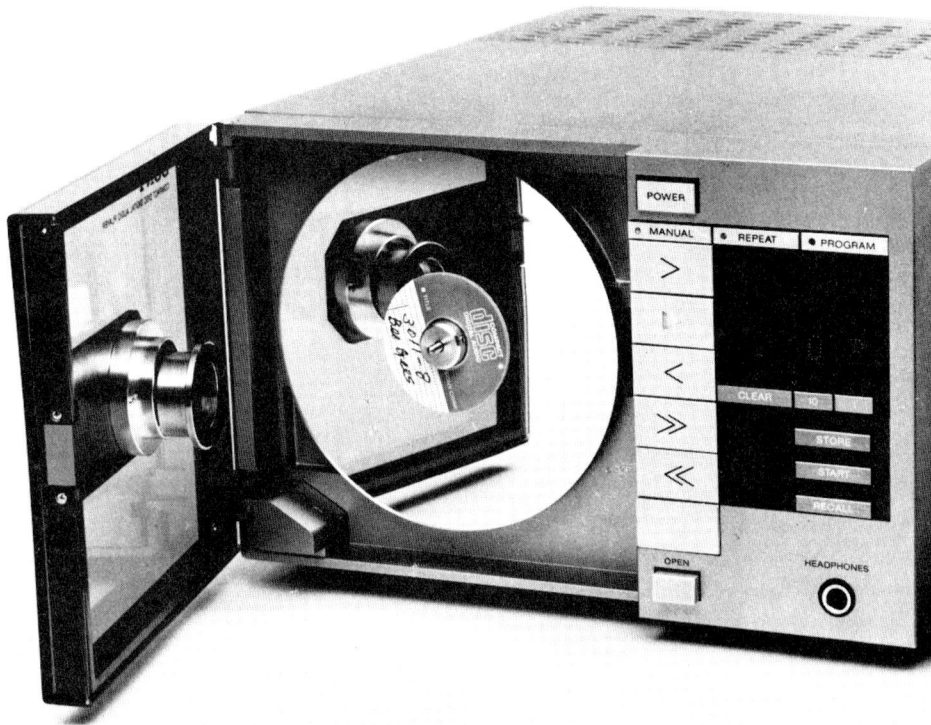

Digital hi-fi components such as this disk player are just becoming commercially available. The player uses a record of about five inches in diameter. The audio information is encoded as a series of pits on the record's surface. A laser beam in the player reads the pits and produces a digital signal that is then converted into sound.

could automatically convert audio signals directly into digital code, without needing human operators. At that time the only switches available were bulky electronic vacuum tubes. Reeves' circuit needed so many switches that it would have filled a house. Eleven years later, the invention of the transistor after World War II made it possible to build a digital coding circuit of a more practicable size. Since the early 1970s engineers all over the world have been building digital audio coding devices and in the future these may replace conventional methods of recording sound.

The advantage of digital audio techniques is that the pulse stream is almost totally immune from distortion and background noise. The replay equipment recreates an analog signal under instruction from the digitally coded signal, so as long as the pulses are still clearly recognizable and the digital playback system does the job correctly, the sound played back will be nearly perfect.

Digital hi-fi is only now appearing on the domestic market because the coding and decoding equipment necessary is very complex and has taken a long time to develop. To make a digital record or tape player of reasonable size and cost engineers have had to integrate many thousands of individual transistors onto a single microchip. Also, the stream of pulses needed for hi-fi stereo sound is so rapid — a few million per second — that it is not possible to record them on a conventional tape or record and so special disks have had to be developed.

In 1979 a Dutch company announced the development of a small record made of light-reflective material about five inches in diameter with digital audio code impressed on the surface in a series of pits. The record is played on a special record player with a laser and light sensor. As the record rotates the pits pass through the laser beam and cause fluctuations in the light reaching the sensor. This produces a digital signal which is decoded to produce sound. In 1980 a Japanese company joined with the Dutch company to develop the system further. Variations of these record players are available to the public today and are being so widely advertised that laser systems will soon be a household word. In the next five to ten years such systems might well replace traditional hi-fi.

Video: Tape

The objective of all video recorders is to capture television sound and picture information as variations in magnetic particles on a moving tape. Shown below are three recording techniques: stationary head (left), quadruplex (right), and helical scan (facing page). Also shown is a cassette recording head (top left).

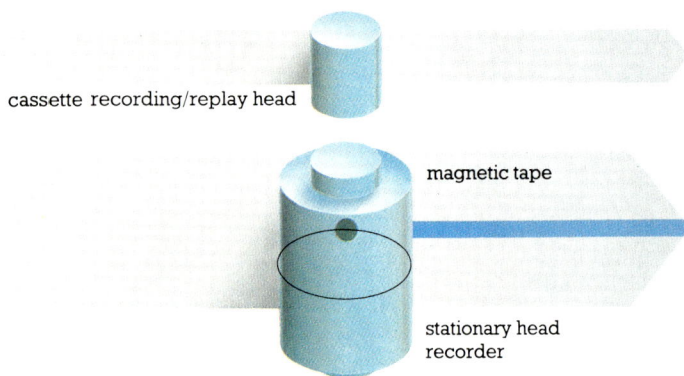

cassette recording/replay head

magnetic tape

stationary head recorder

magnetic particles

quadruplex head recorder

A stationary head recorder works the same way as an audio tape recorder, with a single head recording a track the length of the tape. Because the video signal contains more information than audio, the video tape must move very fast past the recording head.

Sound can be recorded on tape as a changing pattern of magnetic signals, and the same can be done with television pictures. How? With video. Video signals, especially those needed to produce color television pictures, cover a wide band of frequencies. If you try to record video on an audio tape recorder the individual waves of the signal merge together in an indecipherable jumble of sounds. Because you are recording two separate and different wavelengths— one for sound and one for image — the amount of tape surface that passes the recording head in one second must be able to store up to 20,000 separate audio wave cycles, and several million separate video cycles. The faster the tape moves past the heads, the more tape there is per second on which to store the wave cycles.

There are two ways to increase the amount of tape surface passing the recording head. One is to run the tape past the recording head at high speed, similar to speeding up an audio tape recorder. The other is to allow the tape to move at the same speed used for normal audio recording and to spin the magnetic head rapidly so that it scans a series of lines across the tape width. All professional and commercial video recorders today use this rotary head scan technique to obtain high-quality recordings.

The first rotating scan recorders were developed

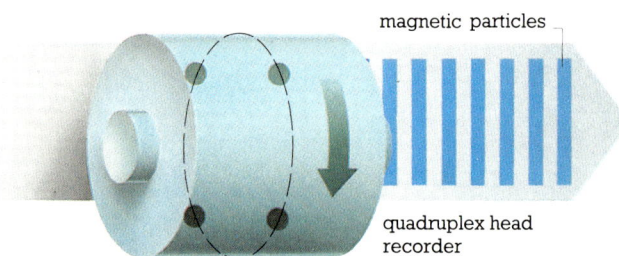

With a quadruplex recorder, the tape moves at a more manageable speed of fifteen inches per second. While the tape moves, the four heads themselves rotate, recording tracks almost at right angles across the width of the tape.

in the mid-1950s. Wide magnetic tape was run at audio speed — fifteen inches per second — past four recording heads mounted on a drum. The drum rotated so that the heads crossed the tape repeatedly at nearly right angles to its direction of travel. The result was a length of tape with a magnetic recording laid across its width in stripes, like the rungs of a ladder. The system was called quadruplex because of the four heads on the drum. Initially such methods could only produce poor quality black and white pictures; but the system was soon improved to provide sharp pictures in full color. It became a world standard for broadcasters and is still in use today.

Most present-day domestic and professional recorders work on what is called the helical scan principle. In a basic helical scan recorder two heads are mounted on a drum that rotates alongside the tape. Whereas the quadruplex drum heads sweep across the tape at right angles, the heads of a helical scan drum sweep obliquely. A helical scan recording thus has a series of magnetic tracks laid obliquely across the tape width like the threads of a screw. The obliquely laid tracks are long enough to record half of a complete television frame, and it is possible to show a single frame on a helical scan recorder simply by stopping the tape. The two heads are thus kept scanning the same magnetic tracks over and over again.

With the advent of video cassette recorders, domestic video recording became a commercial reality. Because it is difficult to lace the tape of a helical scan recorder around a rotating drum, the tape in a video cassette is automatically threaded around the drum by a mechanical arm inside the recorder.

The first domestic video cassette recorder was produced in the early 1970s. The early domestic machines were expensive and could only record one hour of programs on each cassette. Engineers have solved the second problem with a technique called slant azimuth. Two heads on a drum are fixed at slightly different angles so that the tracks they lay down on the tape are slightly different. Electronic circuits ensure that each head plays back only its own tracks. This enables all the tracks to be packed closely together across the tape without interfering with each other, which in turn means that less tape is required to record a given amount of programs.

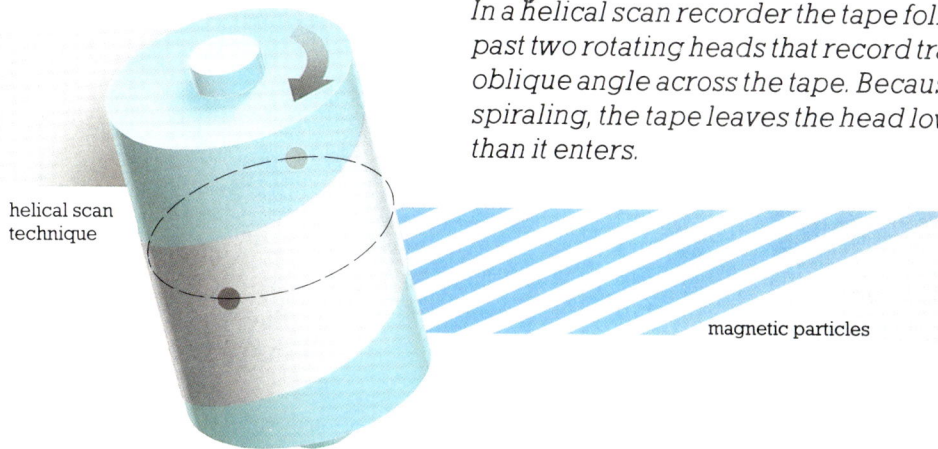

In a helical scan recorder the tape follows a spiral path past two rotating heads that record tracks at an oblique angle across the tape. Because of the spiraling, the tape leaves the head lower than it enters.

helical scan technique

magnetic particles

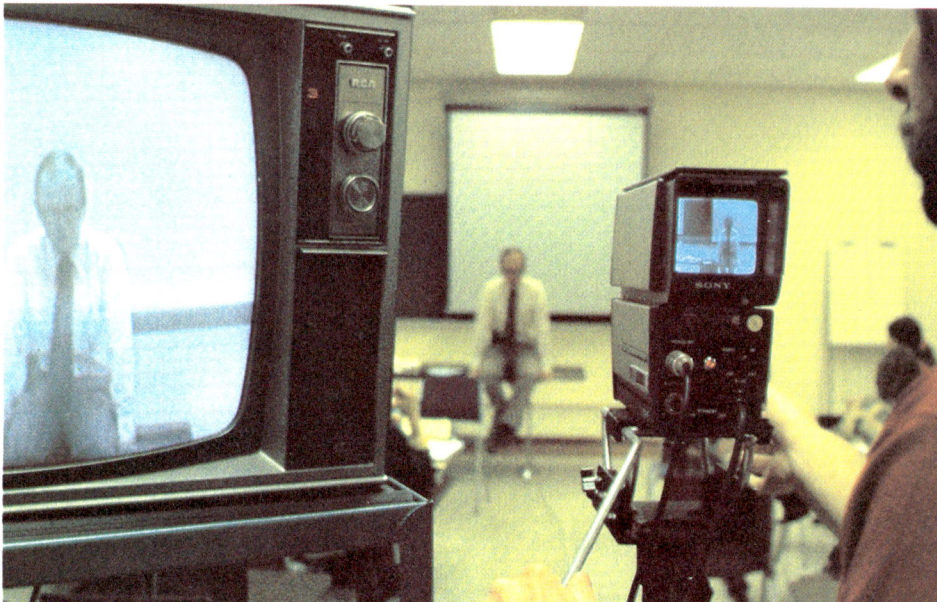

In a helical scan recorder, mechanical arms pull tape around a recording cylinder.

The cameraman at left is watching a monitor on a camera and televison screen.

Video: Disks

It is difficult to record moving images from television cameras onto a disk for the same reason that it is difficult to put video onto tape: television's millions of wave cycles per second simply run into one another. A video disk must be able to record analog signals of six million cycles per second, or digital pulses of seventy million per second. The solution to this problem, used by all video disk systems, is to increase the speed at which the disk turns so that more of the disk's track passes under the tracking head in a given period of time.

Another problem is the difficulty of recording high-frequency signals on a disk made in the conventional way. For these reasons engineers have had to explore and create new ways of recording video signals onto disks.

A precursor of today's video disk system was developed in 1928 by John Logie Baird, the creator of the first mechanical television system. In his invention, Baird used an ordinary shellac record to record still black and white pictures to use on his mechanical television. He was able to do this because his television needed only low-frequency signals.

In the 1970s a vastly improved version of the Baird idea was developed. A tiny stylus (or needle) tracked a super-fine groove in the surface of a flimsy disk which rotated at high speeds. The system was able to produce color pictures and was launched commercially in the mid-1970s. Bold though this venture was, it failed to be commercially successful because the disk's playing time was too limited.

There are now three different types of video disks that should be available to the public in the 1980s. The three types are an optical disk, a grooved disk,

In this photograph, a laser beam focuses onto the pits of a video disk magnified 10,000 times. The pits are created by the pulses governing the scanning beam in a television. When played, the laser reads each pit, converting them into electric signals.

A laser beam focuses on a video disk moving at up to 18,000rpm. Passing through the beam, a light-sensitive cell picks up variations in the pits and converts them into electric signals.

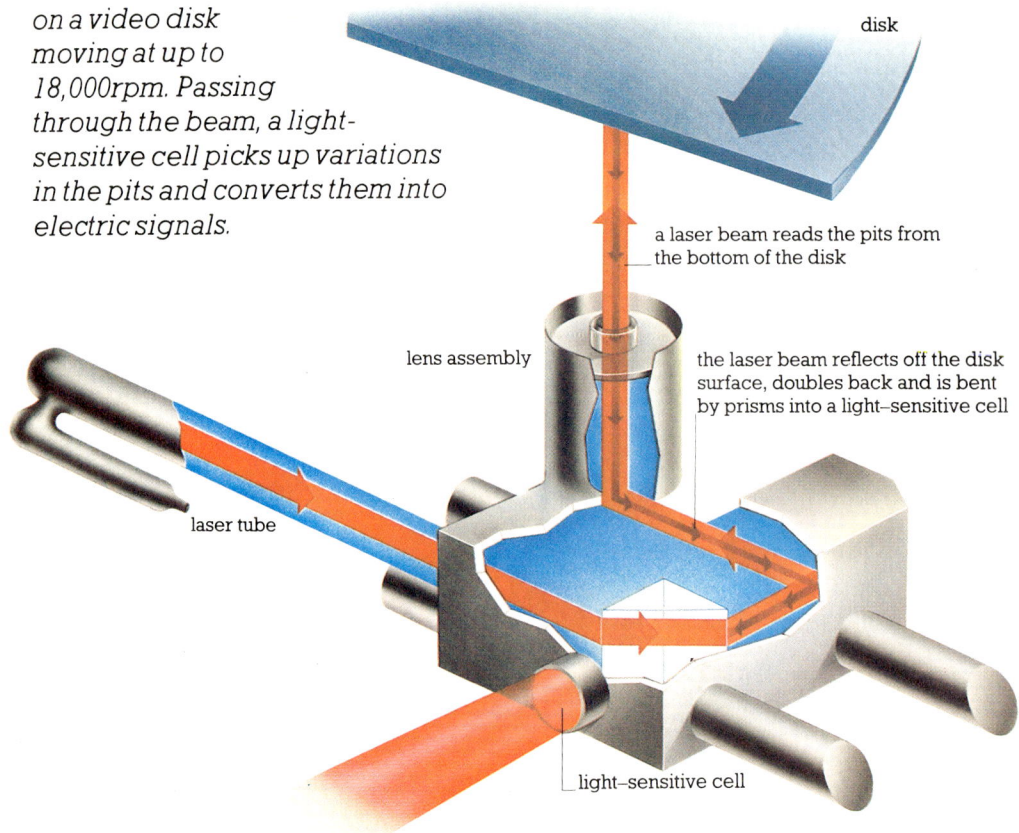

disk

a laser beam reads the pits from the bottom of the disk

lens assembly

the laser beam reflects off the disk surface, doubles back and is bent by prisms into a light–sensitive cell

laser tube

light–sensitive cell

The surface of a laser disk is covered with millions of tiny pits. These tiny pits break up natural light into its component colors to produce the rainbow effect seen in this picture.

This enlarged view of the stylus and grooves of a video disk system show how the tiny pits on the bottom of the disk's groove are sensed by an electrically conducting stylus.

In this grooveless system, electric signals are produced by a stylus as it passes over the pits. This keeps the stylus aligned and eliminates the need for grooves as in the other system.

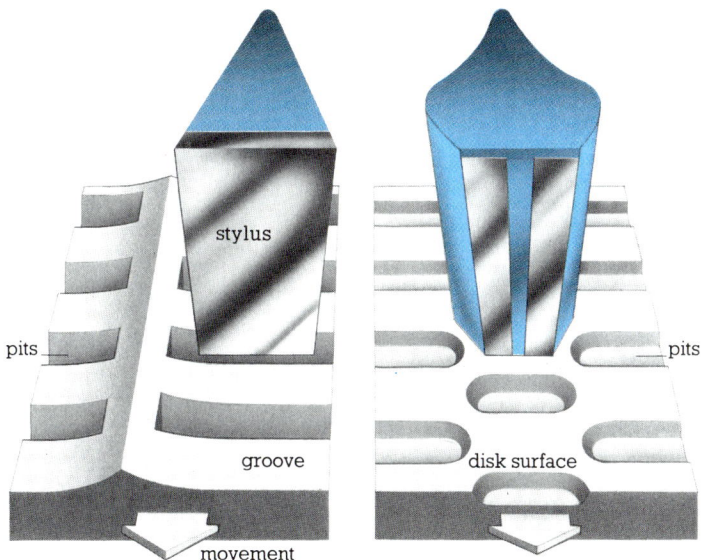

stylus

pits

groove

movement

pits

disk surface

and a grooveless disk. Obviously it is too soon to tell which of the three systems will provide better quality at less cost to the consumer.

In the optical version a pencil-thin laser beam shines onto the surface of a disk that looks like an ordinary record but has a track of intermittent microscopic pits instead of grooves in the surface. As the disk rotates, the laser senses the pits and converts the pattern and intervals — arranged in a code — into a television signal. One advantage of this system is that the disk will not wear out quickly because there is no physical contact with the disk.

The grooved system uses a grooved disk that looks similar to an ordinary LP. The grooves, however, are much finer and are tracked by an equally fine stylus. The walls of the grooves are smooth rather than undulating, as on a regular LP. The disk itself is made from a special plastic that conducts electricity (most plastics do not). Microscopic pits in the bottom of the disk groove produce electric signals in the stylus, which contains a tiny electrode. These signals are amplified and fed into a television set to be displayed as color pictures and sound.

The grooveless system also uses a conductive record and electrode stylus but there are no grooves on the disk's surface. Instead, the stylus is flat and skims across the surface of the disk, guided by electric signals generated by the pits, which also carry the picture signals.

Grooved and grooveless disks are easier to manufacture than the optical disk, but they suffer from the disadvantage that the electrically conductive material on the disk is easily damaged, if, say, it is touched by a hand or picks up dust. Therefore, both are kept in a protective cover and are automatically handled by the video machine itself.

All video disk systems available to the public — or soon to be made available — can replay prerecorded disks, and in the future video enthusiasts may be able to record their own programs. The record could be a version of an optical system in which the laser burns the pits that carry the recording, or a new idea currently intriguing video engineers — a magnetic disk.

83

Network Systems

The "wired society"— a term heard more and more frequently — refers to a possible revolution that experts are predicting will result from advances in electronic communications. According to these experts, commerce, politics, emergency services, education and entertainment may all soon rely on an electronic network, linking all television sets in a community to a central computer.

Already a few places in the world are equipped with trial networks that may be forerunners of the wired society. The most extensive of these experiments so far is a system called Qube (pronounced "cube") introduced in 1977 in Columbus, Ohio, by a cable television company. Qube is a combination of cable television, viewdata, and an emergency services network available to subscribers for a monthly fee. With Qube, an ordinary television set used with a small, hand-held computerized terminal is linked through television cables to a large, central computer. This computer monitors each home terminal every six seconds to make sure equipment is functioning properly; to check which of the thirty available channels the subscriber is watching; and to find out what responses, if any, the subscriber is sending back to the main computer.

In addition to a full range of cable television programs, subscribers can get the daily newspaper and constantly updated stock market reports or the weather. They can play video games and even take college exams without leaving their living rooms. Some banking functions can also be performed through the Qube program; for example, bills can be paid, and loan applications filed.

A subscriber at home can also do some shopping through Qube. Local department stores provide programs of their available merchandise that can be called up on the television screen. Orders are placed through the consumer's terminal.

Emergency connections to fire departments, police, or local hospitals are also possible with Qube. Electronic heat, smoke, and burglar sensors in the home are connected to the central computer. If a sensor is activated, the computer alerts the appropriate

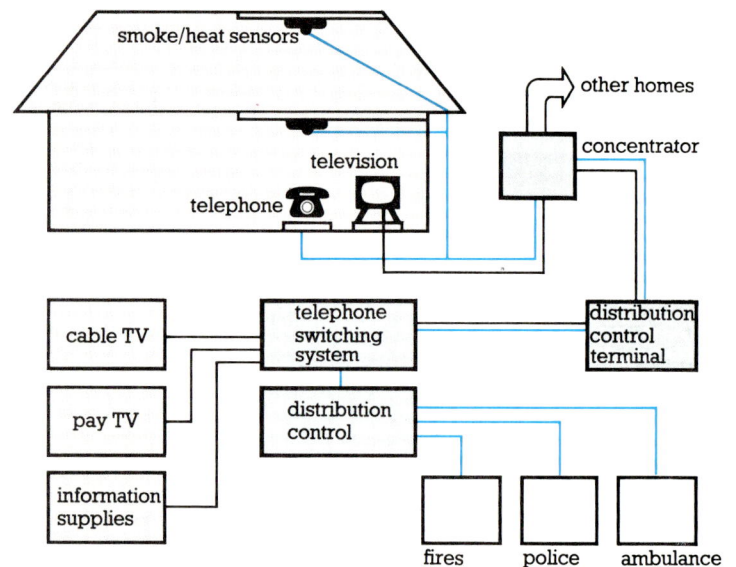

In the near future, many homes will use networks such as the one shown here. The black lines indicate information supplies and the blue lines emergency sensor services.

service and instantly provides them with important data. In the case of fire, for example, the computer gives firemen the location of the nearest hydrants, and a description of the house — including the location of rooms in which children or invalids might be sleeping. For elderly or chronically ill persons the system offers special emergency services. A person can call an ambulance even if he cannot reach a phone by using a piece of jewelry that sends a remote signal through the home terminal. As the ambulance is dispatched, attendants receive computer printouts containing such information as the patient's history and the name of his doctor.

Qube can also be used for electronic town meetings in a "wired" community. The town officials would appear on television, present the issues, and votes would be sent through the home terminals and counted by the central computer. So far, Qube subscribers have not been able to vote officially, but they have expressed their views on a number of points.

It is not yet known whether enough people will

This family is enjoying the services of a network system. In some cases such a system will allow a family to receive cable TV, regular TV, viewdata and teletext to provide a continuous source of entertainment, news, weather, and many other kinds of useful information.

The system at left is part of a distributed network that enables one computer to talk to and work with many other computers over telephone lines. This system can convert spoken commands into its own language.

want the Qube services to make it a profitable enterprise. Nevertheless, the experiment has already spread to five other American cities and comparable systems have been installed in other parts of the world. In the western Canadian city of Regina, Saskatchewan, an electronic system called Telidon is being tested in a number of homes. A model town outside Osaka, Japan, is trying its own version, with the system linked by fiber optic cables rather than conventional telephone or television cables. The signal-carrying capacity of the fiber optic cables offers the possibility of many more services than the other network systems.

Two-way electronic systems, however, could be susceptible to abuse. For example, unscrupulous operators could monitor the services a subscriber uses, record his viewing habits, and sell this information to advertisers or other interested parties. But before these networks become widespread, technological and legal safeguards will probably be found to ensure an individual's privacy.

Electronic Office

Electronic office equipment, such as the word processors (left), telex machine (right), and computerized telephone reception console (foreground), is making communications between offices more efficient than ever before.

One of the largest markets for communications technology is the office. This is partly because an increasing number of people now work in offices, and partly because of the mass of information that circulates within and between offices. New applications for computerized communications systems can greatly increase efficiency by eliminating tedious tasks and speeding the flow of information.

One of the most time-consuming office tasks is the preparation of letters and internal memos. A computer that can do this job is called a word processor, now a common sight in many offices. Word processors are electronic typewriters that have a display screen and small computer memory. A person types the letter or message on a keyboard and simultaneously sees on a display screen what is being typed. The operator can correct errors by deleting or adding letters or words, or by moving sentences and paragraphs around. Once the letter is ready, the operator merely pushes another key and the word processor rapidly produces a perfectly typed letter. A word processor can store standard letters and hundreds of names and addresses in its memory so that "personalized" form letters can be produced in very little time. Another machine puts the letters into envelopes and stamps them.

In the electronic office of the future even this method of preparing and sending letters will seem old-fashioned because letters will be sent without using paper. A businessman, for example, will type a message into his own computer terminal and send it to a colleague in the next office — or to somebody on the other side of the world over a telephone line.

Many offices now keep their files on magnetic tape or magnetic disk. The latest photocopiers can copy a specially prepared piece of paper onto a display screen or automatically file it with other computerized information.

The brain of the electronic office is the computer,

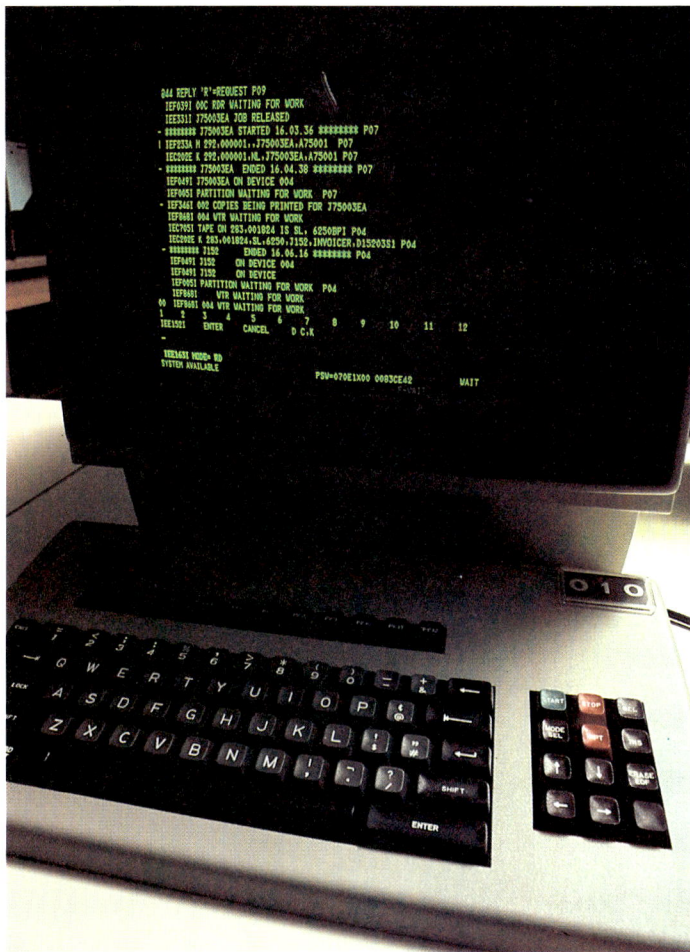

A word processor terminal, keyboard and display screen such as this can store information that can be updated regularly. The unit here is used by an engineering company; the image shows the status of jobs in progress.

At right and below are three configurations used in local area networks that can link offices in one building or office complex. Each office can be linked directly, as with the ring or bus method, or through a central switching system, as with the star method.

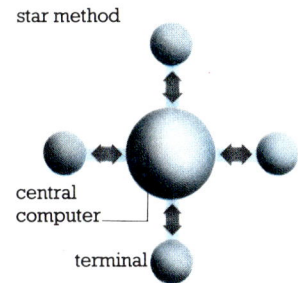

star method

central computer

terminal

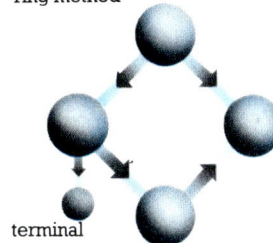

ring method

terminal

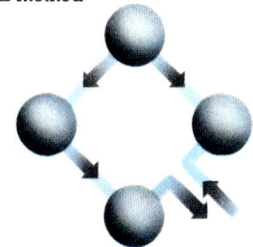

bus method

and the communication links are its nervous system. As electronics technology advances, computers have become more powerful, smaller, and cheaper. The smallest businesses can now afford a microcomputer; by 1990 no business will survive without one.

While most small office computers can operate on their own, the electronic office is most efficient when its computers can communicate with other computers. Long-distance communications can go through the telephone system, but for local communications — such as those between offices in one building or industrial park — the so-called local area network has advantages over conventional telephone links. Local area networks offer direct connection between all the machines in the network and can transmit large amounts of data at a low cost because of the short distance between machines. They allow small office machines not only to exchange data among themselves, but also to borrow the services of expen-

sive, high-capacity computers. Networks can also form the basis of an in-house electronic mail service.

There are three arrangements for communications networks between office machines. One, the star configuration, is widely used for long distance communications and conventional local communications networks. This has a private business switching center with cables radiating out to each machine.

For local area networks, the most popular configurations are the ring and bus. In the ring network each machine is connected onto a ring of cables. Messages are sent in electronic packages each with an address. When the message reaches the destination, say, halfway around the ring, a machine recognizes the address, reads the message and tells the sender that it has been received. The message then goes to the relevant person's terminal. In the bus network all of the machines are linked together by an open-ended cable.

Voice Recognition

The most common form of communication between people is speech. The human brain is capable of deciphering the meaning of a wide variety of sounds and understands different accents and intonations. This unique ability of the human brain has, so far, outstripped all manmade machines. Until there are very sophisticated versions of what engineers call voice recognition machines, the job of putting information into a computer will be cumbersome and slow. Engineers have built machines that can add much faster than the human brain and store an endless quantity of information. There are even machines that can compose music and design and make other machines. But computers capable of recognizing and understanding human speech are still in the early stages of development.

A computer has little difficulty understanding mathematical logic or the logic of design. These are manmade theories that can be explained to the computer through programming. But the human voice is extremely irregular and does not follow logical patterns. Even a scientist cannot fully explain how the human brain — which is a highly intelligent and complex machine — recognizes speech. Because of this disadvantage, scientists have had to work with very little information in their attempts to develop voice recognition systems.

In addition to the irregularity of human speech, a voice recognition computer also has to cope with varying pitches of sound — the differences between a man's and a woman's voice, for example. There are also the problems of background noise in the

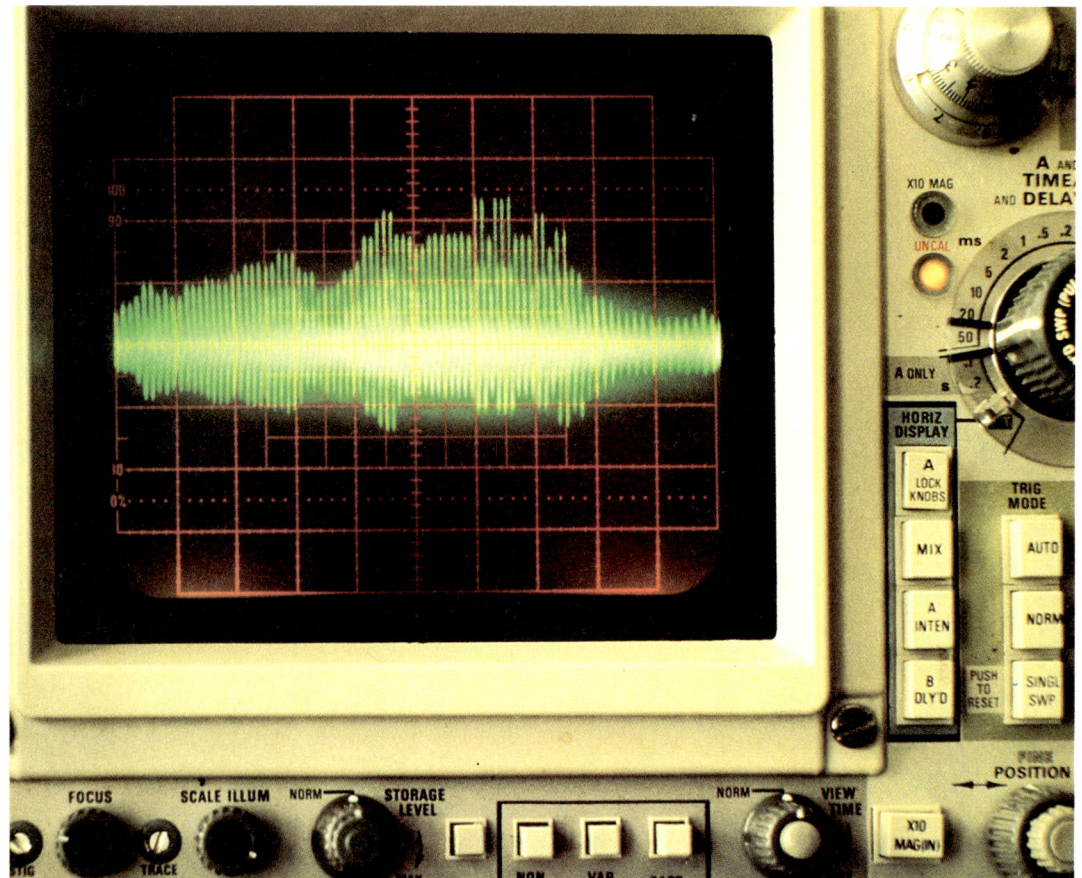

Part of the sentence "Where are you?" appears on a display screen of an oscilloscope as a series of evenly spaced lines of varying height. Part of the technique of voice recognition involves converting sound waves into digital electronic signals.

Computers are now being used to help people with speech defects. In this picture a deaf child is being taught to speak by watching the pattern of his voice sounds on a computer screen.

speaker's environment as well as interference and signal fluctuations within the computer itself.

Two approaches have been tried. Computer scientists can train a computer to recognize the voices of a few people using a limited vocabulary — or they can train a computer to understand a smaller number of words, regardless of who speaks them. One method of extending a computer's memory of words is to teach it sets of vocabulary, each triggered by a key word. If, for example, voice recognition were used to check stock in a store, the key word might be 'stationery'. The computer would then exclude all the words in its vocabulary that were not related to that word and focus only on those that were.

In developing a voice recognition system the computer scientist breaks down sentences into units that the computer can analyze. This can cause problems when words flow into each other, but it also means that the machine is able to recognize several voices, even though they may contain different sounds.

If a computer is to understand a continuous flow of speech, sentences must be broken into even smaller units called phonemes, or the sounds that make up words. The computer then learns to recognize a word as a group of phonemes. When it has worked out the patterns of certain sounds it compares them with its own memory of patterns and selects the word that resembles it most closely. As it builds a phrase the computer can tell from the structure of a sentence which word is most likely to be correct. This may seem like a hit-or-miss method, but it may be the way the human brain does the same job.

Voice recognition can be used for putting information into a computer, retrieving it, and for giving instructions. Examples include not only stocktaking in stores, but also the continous updating of information for navigation, and many other tasks where it is not possible or convenient to use a keyboard to obtain information. In addition, speech recognition could also be used in factories for giving machines instructions, or by disabled persons to switch appliances on and off. Once perfected, it could pave the way for many useful applications.

This is a voice recognition module—a tiny device that can recognize up to 100 words or short phrases in any language and recognize spoken words with ninety-nine per cent accuracy.

Telemetry

If you put an electric fuel gauge in a radio-controlled model airplane and used its signal to tell you when it was time to bring the plane down, you would be using telemetry. Telemetry is a means of making measurements and transmitting them from a remote spot to a place where they can be analyzed. For example, it is the technique used by engineers to check and investigate the performance of equipment in service. Knowing about events as they happen can be vital for engineers and researchers who might need to modify the course of an experiment or, if something goes wrong, to follow the sequence of events that led to the point of failure. The uses of telemetry are vast: they range from following the flight of a missile to its point of impact with a target to monitoring the normal operations inside a car that are hidden from view, such as the engine or brake system.

Simple telemetry links work in one direction only. Sensors, or transducers, collect information from several points which is then fed into a central collection site for transmission as an electric signal or by radio waves back to base. There are many types of sensors. Each is designed to relay specific information, for example, the position of a lever, a gas pressure, a temperature, or a voltage.

Telemetry data transmissions are made of many digital bits, coded as electronic pulses. At the site where the data is collected — say on the front of a missile — sensors will produce analog signals that describe some aspect of the missile's performance. These are converted into digital form at the central collection site. Before transmission, the telemetry equipment puts all the digital information into a predetermined sequence so the receivers will associate each set of digitized data with the proper sensors. The information is then fed down a telemetry link or through a highly efficient communication channel to an observer or operator on the ground.

The digital bits are recorded at the receiving end for later use or they may activate machines that plot graphs or display the transmitted information as different levels on dials. Engineers use telemetry during flights of new aircraft, for example, to reconstruct the readings of cockpit instruments electronically. This allows an observer on the ground to read the cockpit instruments at the same instant as the crew in the airplane, so that pilots on the ground can monitor

This view of the space shuttle Columbia's cargo bay was taken during a door-opening-and-closing exercise on the first day in flight. The device extending to the right is a remote cargo arm used for removing satellites from within the ship. Such functions are carried out by using telemetry equipment on earth.

safety actions and engineers can watch technical failures as they happen. Such tests are also used on car engines where engineers can watch graphs and dial indications to tell them exactly what is happening. Telemetry is also used in satellite systems, all kinds of military hardware, and such dangerous situations as nuclear power stations and chemical processing plants.

More complex two-way telemetry is used by engineers to control experiments in progress. In tests of new aircraft, for example, just as the radio enables observers on the ground to talk regularly to the crew in the plane, in some systems the telemetry link itself feeds information from the test center to the experiment. During a ground test, for example, the engineers might send signals simulating the failure of vital components to watch how well the aircraft's back-up systems cope in an emergency.

The most sophisticated telemetry systems are fitted to spacecraft. Voyagers' trips past Jupiter and Saturn were not a simple operation. Hundreds of men and women — their eyes fixed on television screens for hundreds of hours — monitored every fraction of each spacecraft's sophisticated equipment, its power levels and camera angles. Continuous adjustments were required to keep the spacecraft functioning at optimum performance. The ground personnel, and the success of the missions, relied almost entirely on the technology of telemetry, without which none of this would have been possible.

The flight deck of the space shuttle Columbia. In front of the flight computer and navigation aids console are three cathode ray tubes used to display computer information for the crew. The commander and pilot each have a rotational hand controller in front of their seats. These provide electrical information to the computer system, which then controls commands for altitude changes.

The Incredible Future

It is the winter of 1999 and you are bored. The weather is cold and you have been trying to entertain yourself by watching one of the 300 three-dimensional television programs that are being beamed to you via satellite from all over the world and from special stations in orbit around the earth. You have tried all of the games, competitions and offers on the cable television links into your computer television screen and you have even written a letter and mailed it on the electronic mailing device — but for some reason you are still bored.

You decide to get in touch with a friend. You press a few buttons, the television screen in front of you switches on, and a computer tells you that it is tele-phoning your friend. In a few seconds the screen fills with a live picture of his face and his voice comes out clearly and loudly asking if you would like to go ice-skating with him. Still in the voice mode, you tell the computer to order a taxi. It replies by telling you that one will be at your front door in a few minutes.

Is any of this possible, or will it be possible in the not-so-distant future? The technological details that would make this scene a reality are all feasible — and some are almost around the corner. Television programs are already beamed directly into private homes by satellite and may be widespread in the next few years. Such transmissions can improve picture quality and make programs available to more people than ever before. Although still in the laboratory stage, three-dimensional television programs can be transmitted today and received on special television sets. There is no technical reason why such programs cannot be beamed to a local television cable station from anywhere in the world. Communications from space are routine for military and exploration missions — they could just as easily become so for domestic and commercial use.

As components become smaller, it is likely that your home computer will be built into your television set. One day, you will be able to have a flat-screen television in every room of the house, simultaneously broadcasting the same shows. Or, you could tell the screen to display pictures or images of your choice.

You could ask for computer-generated pictures of famous paintings, for example, or simply a pleasant landscape to brighten a bare wall.

There is just one technology described that is beyond the reach of communications engineers — the ability of a computer to recognize, interpret, and respond in fairly complex speech. But this is a subject of concerted research today and may one day be commonplace — with the unrelenting advances in microelectronics, anything seems possible.

Even now, in fact, research scientists are predicting living microcircuits so small that the individual components will be no bigger than a molecule. All living matter responds to or produces tiny electrical currents, so why not direct them to novel use? Some scientists already believe they will be able to implant biochips into people's bodies, initially to correct defects such as wasted muscle or malfunctioning organs. Engineers may soon be designing biochips that can be implanted into paraplegics to give them fuller use of their muscles. Already in practice is a tiny microprocessor that is used to correct and monitor a patient's heart rate. Similar to the conventional pacemaker, the microprocessor can regulate a person's heart beat as well as continuously store information. Thus the patient's physician can find out what is happening inside his patient's body at that moment, or what has been happening over a period of time. The step from implanting a microprocessor in a computer to implanting a biochip in a human being is a relatively small one.

Huge satellites such as the one shown here may play a major role in the future of communications to transmit information about space to earth and between different places on the globe, and to communicate with one another, much as computers do on earth today.

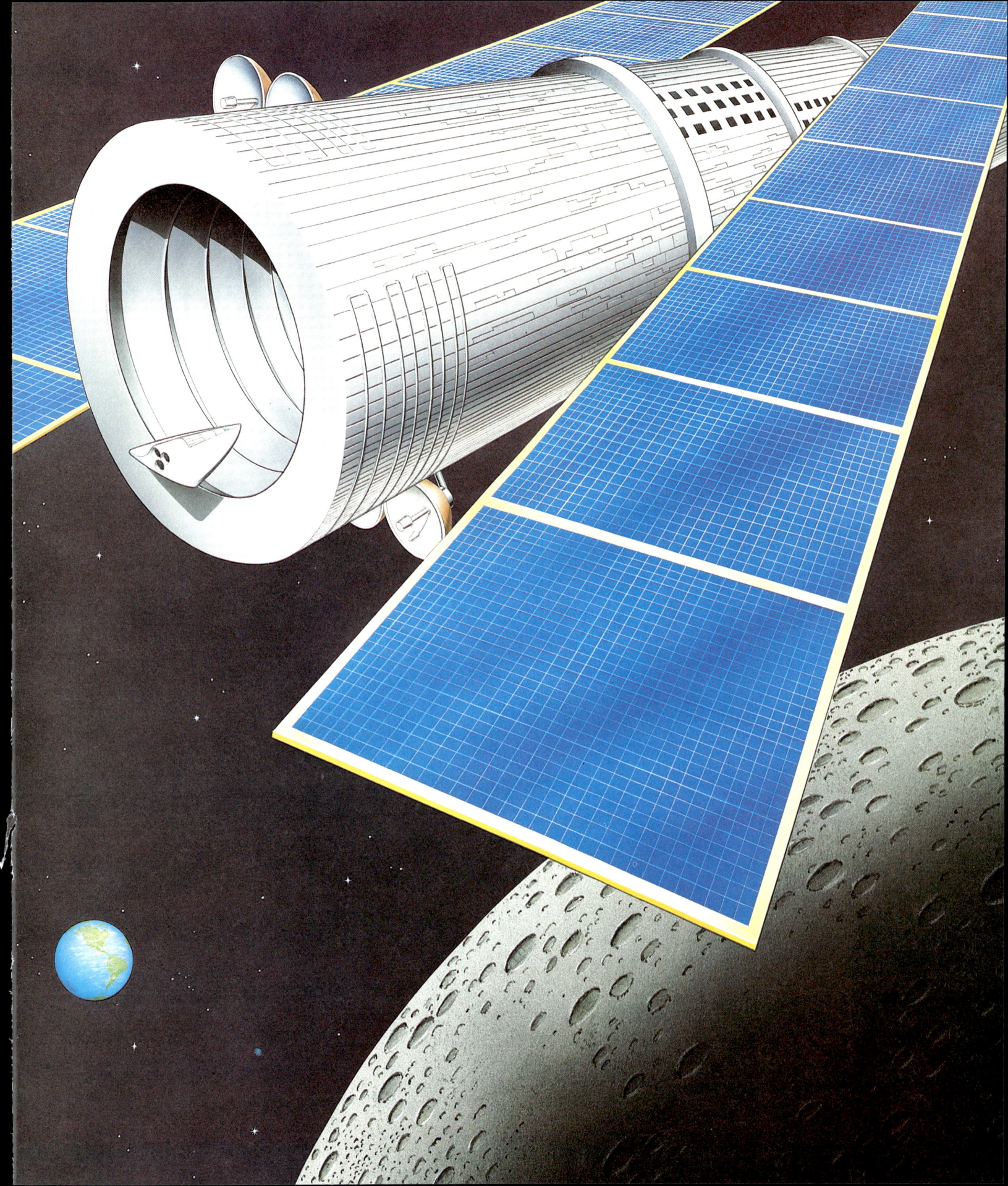

Glossary

Amplification: an increase in the strength of a signal

Amplitude modulation: varying the amplitude (height) of a carrier wave in accordance with information to be transmitted

Analog signal: electrical signal that varies directly with the information it carries

Antenna: a metallic structure that captures and transmits electromagnetic radiation

Atom: the smallest part of an element that can take part in a chemical reaction

Binaural reproduction: a method of recording sound precisely as a person would hear it for replay through earphones

Bit: a contraction of the words "*bi*nary di*git*". A binary digit is either one or zero. One is represented electronically as a precisely timed pulse of electricity, zero as a precisely timed pause

Cable: a group of conductors carrying electrical or lightwave signals

Cathode ray tube: the heart of a television set or visual display unit consisting of one or more electron guns that direct an electron beam onto a phosphor-covered screen

Chrominance signal: the part of a television broadcast signal that contains the color information

Circuit: an arrangement of electronic devices to conduct electrical signals or voltages

Circuit switching: a technique for sending information in which a circuit directly connects the sender and receiver

Citizens band: a spread of frequencies available to private citizens for transmitting and receiving radio signals

Coaxial cable: a cable with inner and outer conductors insulated from each other by plastic disks

Conductor: a material through which electrons flow

Continuous wave: an unbroken oscillating electrical signal

Crossbar switch: telephone switches with relays arranged in a grid

Current: a stream of electrons through a conducting material

Definition: the quality of a television picture in terms of the smallness of objects that can be seen

Digital signal: electrical signal composed of a stream of pulses

Diode: an electronic switch in which current can move in one direction only

Doppler effect: the apparent change in the frequency of a sound or signal caused when the source of the sound or signal moves towards or away from the receiver

EHF: extremely high-frequency radio waves

Electrode: component to transmit electricity into or out of a gas or other fluid

Electron: negatively charged body that is part of an atom

Electron gun: device that produces a stream of electrons

Electromagnet: device exhibiting magnetism when electricity is connected

Facsimile: machine that transmits text or images over telephone wires

Frame: one television picture. In the United States, sets show thirty frames per second.

Frequency: in a wave, the number of complete cycles per second

Frequency-hopping: rapidly switching frequencies during radio transmission to keep eavesdroppers from listening in

Gateway: a system that enables one computer to gain access to the data of another through a public network

Geosynchronous orbit: orbit in which a satellite, if it could be seen from the earth, would appear to be stationary; geostationary orbit

Germanium: semiconductor element

HF: high-frequency radio waves

Hole: space left by an electron that

has moved from its place in a semiconductor crystal

Integrated circuit: a circuit in which the transistors and other components have been formed on a small wafer of one material, normally silicon

Jamming: preventing a radio receiver from tuning in to a particular signal

Laser: device that produces light in a highly focused beam

LF: low-frequency radio waves

MF: medium-frequency radio waves

Optical fiber: thin glass fiber through which light signals travel

Packet-switching: technique for sending information by grouping data into packages

Phosphors: chemicals that emit light when hit by electrons

Photons: the basic particles of light

Pulse code modulation: a technique for converting an analog signal into a digital signal

Pulse radar: radar in which energy is emitted in high-powered bursts

Radar: radio-based device used to detect objects too far away to be seen by the naked eye

Raster: a grid of horizontal lines; the pattern in which an electron beam scans a television screen

Receiver: device that converts radio or television signals into sound and pictures

Repeater: strengthens a telephone signal that has weakened down a line

Scrambler: device that mixes up a voice on a telephone so that eavesdroppers cannot understand

Semiconductor: material whose conductivity falls between that of an insulator and conductor

SHF: superhigh-frequency radio waves

Shortwave: now called high-frequency signals

Silicon chip: a wafer of silicon that contains integrated circuits

Solar cells: electronic circuits that convert sunlight into electricity

Solid-state: relating to the electronic properties of solid materials, especially semiconductors

Sonar: sound-based system for detecting objects underwater

Stereo: equipment that transmits sound in two channels, each reproducing the sound from a different direction

Switch: device that can be controlled to pass electricity or not

Switching system: used to connect telephone calls

Teletext: information service that

can be received on a modified television

Triode: vacuum tube with three electrodes

Transistor: a semiconductor device that can switch or amplify a large current flowing between two internal regions by a small current applied to an intermediate region

Transmitter: device that sends signals down a wire or into an antenna

Troposcatter: the effect of the troposphere on UHF signals

UHF: ultrahigh-frequency radio waves

Vacuum tube: a device used to produce and control a stream of electrons; basic amplifier in early electronic systems

VHF: very-high-frequency radio waves

Viewdata: system that connects a television set to a computer store of data over a telephone line

VLF: very-low-frequency radio waves

Waveguide: a device that traps electromagnetic waves internally so that they can be directed from one destination to another

Wordprocessor: combination of computer and typewriter with a display screen dedicated to the efficient production of text, tables, and graphics

Index

antennas 28–9
  signal frequency 28
  types 28

Bell, Alexander Graham 6,
  10, 76–7
binaural sound 77

citizens band radio 31
computers 62–3, 68–9
  circuit-switched networks
    68–9
  connecting to telephone 68
  ESS telephone systems 62
  information revolution 62–3
  packet-switched networks
    69

digital communications 14–17,
  46–7, 50–1, 60–9, 78–9,
  80–3
disks 70–1, 82–3

Edison, Thomas 52, 70
electricity 8–9
electromagnetic waves 9
electronics 52–5
  diode 53
  "Edison effect" 52
  office 86–7
  integrated circuits 55
  valves 52–3
  solid-state transistor
    54–5
  triode 53

facsimile 20–1
fiber optics 48–9
frequency-hopping 35

hi-fi, digital 78–9
hi-fi, traditional 74–5
  amplifier 74
  feedback loop 74
  loudspeakers 74
  turntable 74

integrated optics 50–1

light 46–7
  early uses 46–7
  electric 47
  fiber optics 48–9
  integrated optics 50–1

Marconi, Guglielmo 6, 24
microchip 26–7, 55

microwaves 32–3
  uses 32–3

network systems 84–5
  office 87

office equipment 86–7
optics 48–9, 50–1

quadraphonics 77

radar 36–7
  continuous wave 36–7
  pulse 36
  uses 36–7

radio 24–7, 30–1
  amateur and mobile 30–1
  amplitude modulation 24–5
  automatic selective calling
    27
  wavebands 30–1
records 70–1
  Berliner's gramophone 70
  Edison's phonograph 70
  grooveless 71
  rpm and grooves 70–1

satellites, civil 40–1
  geosynchronous orbit 41
  MARISAT 41
  microwave antennas 41
satellites, military 42–3
  navigation 42
  reconnaissance 43
satellites as tools 44–5
  astronomy 45
  mapping 44
  weather 44–5
secure communications 34–5
silicon chip 26–7, 55
sonar 38–9
  active 39
  passive 39
  sonobuoys 39
  uses 38–9
sound
  radio 24–5, 26–7
  digital communications 66–7
  disks 70–1, 82–3
  hi-fi 74–5, 78–9
  stereo 76–7
  tapes 72–3
  voice recognition 88–9
stereo and surround sound
  76–7
supertelex 23

tape recorders 72–3
  bias signal 73
  recording and replay 72–3
  tape width and speed 73

telemetry 90–1
  simple 90
  spacecraft systems 91
  two-way 91
  uses 90–1
telephone 10–11
  analog 10
  digital 10, 66
  modem 68
  scrambling 34
telephone networks 16
  long-distance 16
telephone switching 12–15, 66
  analog and digital 14, 15, 66
  crossbar 12–13
  electronic switching 14–15,
    62
  pulse code modulation 14
  solid-state 15
  Strowger 12
telephone wires and cables
  16, 18–19
  coaxial 19
  fiber optics 16, 19
  frequency division
    multiplexing 19
  repeaters 19
  undersea 16, 19
teletext 23, 64–5
television, black and white
  56–7
  early development 56
  how it works 56–7
  rasters 56–7
television, color 58–9
  basic principles 58
  compatability with black
    and white 59
  types of tubes 58–9
television, flat screen 60–1
telex 22–3
transistor 25–6, 54–5
  early development 25–6
  frequency-hopping 27, 35
  microchip 26–7

video disks 82–3
  grooved 83
  grooveless 83
  optical 83
video tape recorders 80–1
  cassette 81
  helical scan 80–1
  quadruplex 80
  stationary head 80
  viewdata 64–5
voice recognition 88–9
  approaches to 89
  problems 88–9
  uses 89

wordprocessor 86

Credits

The publishers gratefully acknowledge permission to reproduce the following illustrations: Alden Electronic Company Inc. 21t; David Baker/ 43; Courtesy of Bell Laboratories 15b, 49l, 50; Paul Brierley 3, 13, 14, 17r, 23r, 53, 58, 70, 73, 81r, 82, 83, 88; British Telecom 11, 12, 19b, 21b, 23l, 64, Bundesarchiv 34r; Cable & Wireless Ltd. 19t, 33; Camera Press Ltd. 91; Capital Radio 27l; Civil Aviation Authority 68, 69; Crown Copyright 31, 47; Daily Telegraph Colour Library 76, 77; Ferranti Ltd. 37; Barry Fox 61; General Telephone & Electronics Ltd. 66; Michael Holford 22, 25, 71b; The Image Bank 17l, 55, 62, 63b, 87; International Computers Ltd. 85b; Interstate Electronic Corporation 89b; ITT Business Systems Ltd. 85b; Interstate Electronic Corporation 89b; ITT Business Systems 86; ITT Consumer Products (UK) Ltd. 65; Brian Johnson 34l; Peter Loughran 71t, 75. 78t; The Marconi Company Ltd. 7, 24, 26; Motorola Ltd. 27r; NASA 90; New Scientist 89t; Pictor International 29, 46, 81l; Plessey Marine 38b; David Redfern Photography 78b; Science Photo Library 15t; Science Photo Library 45 (Earth Satellite Corporation Photography), 40l, 44t (ESA), 44b (Menzel), 40r (NASA), 42 (Polar Record); Tim Sheldon/Photo Graffics 30; Sinclair Research Ltd. 60; Sony UK Ltd. 79; Spectrum Colour Library 36; Standard Telephones & Cables Ltd. 48, 49r; Warner Amex/QUBE 85t; Courtesy of Western Electric Company, New York 32; Westland Helicopters 38t; ZEFA 35, 63t.

Cover photograph: ZEFA.

Artwork by: Gerard Browne 18, 25, 52–53, 54–55, 75, 82, 83, 84; Richard Lewis 8, 9, 10, 20, 30, 37, 39, 48, 61, 80–81; Richard Phipps 13, 16, 29, 41, 52, 54, 64, 67, 69, 71, 72, 87; Ed Stuart 33, 35, 43, 47, 51, 56, 57, 59, 76, 77, 93.

Bibliography

Revolution in Miniature by Ernest Braun and Stuart MacDonald, Cambridge University Press, London, New York.
Getting the Message by J. Barry DuVall, George R. Maughan, Jr., Ernest G. Berger, Davis Publications, Inc., Worcester, Massachusetts.
The Making of the Micro by Christopher Evans, Van Nostrand Rheinhold, New York.
Optical Fibre Communication Systems, C. P. Sandbank, ed., John Wiley & Sons, New York.